WILLIAM COOPER'S A CATALOGUE OF CHYMICALL BOOKS

GARLAND REFERENCE LIBRARY
OF THE HUMANITIES
(VOL. 670)

WILLIAM COOPER'S A CATALOGUE OF CHYMICALL BOOKS, 1673-88
A Verified Edition

Stanton J. Linden

GARLAND PUBLISHING, INC. • NEW YORK & LONDON
1987

Library of Congress Cataloging-in-Publication Data

Cooper, William, fl. 1668–1688.
William Cooper's a catalogue of chymicall books, 1673–88.

(Garland reference library of the humanities ; v. 670)
Bibliography: p.
Includes index.
1. Chemistry—Early works to 1800—Bibliography.
2. Alchemy—Early works to 1800—Bibliography.
3. Magic—Early works to 1800—Bibliography.
4. Medicine—Early works to 1800—Bibliography.
5. Cooper, William, fl. 1668–1688. I. Linden, Stanton J., 1935– II. Title. III. Title: Catalogue of chymicall books, 1673–88. IV. Series.
Z5522.C77 1987 [QD27] 016.54 86-27165
ISBN 0-8240-8557-4 (alk. paper)

Printed on acid-free, 250-year-life paper
Manufactured in the United States of America

CONTENTS

Preface

William Cooper's *Catalogue of Chymicall Books* (1673, 1675, 1688) is a work that is undeservedly neglected by scholars in each of the two "cultures," the humanities and the natural sciences. Though the *Catalogue* is occasionally cited as "the first bibliography of works in English on its subject," it rarely receives more than passing reference in studies of seventeenth-century science and pseudo-science or a footnote in the bibliographies of these subjects. This lack of attention prevails despite the fact that in his own time Cooper occupied a significant place in the London book trade, as author, editor, translator, publisher, book seller, and as England's first book auctioneer.

The purpose of this edition is two-fold. First, it presents a modern, authoritative edition of an extensive list of scientific, medical, and occult books that was last printed nearly three hundred years ago. It thus provides specialists with an important source of authors and titles of books which were newly written (or translated, or re-edited), published, and–perhaps most important in Cooper's mind–available for purchase at his bookshop at the sign of the Pelican in Little Britain, London. *The Catalogue of Chymicall Books* is, then, an excellent guide to the state of scientific and occult book publishing and, by inference, to the audience for such publications in Restoration England. In this edition, the *Catalogue* is supplemented by two appendices: the first consists of additional authors and titles compiled from lists that Cooper included as advertisements in several of the books he published; the second concentrates on his interest in the works of a major contemporary alchemical authority, Eirenaeus Philalethes.

Secondly, the Introduction to this edition presents materials from varied sources, such as the *Stationers' Register*, sale catalogs, and epistles dedicatory, that illuminate several aspects of Cooper's career. These include his achievements as publisher, auctioneer, and author, as well as his strong interest in alchemy and hermetic philosophy. I have subtitled the Introduction "Science and Occultism in Restoration England" because for Cooper and his age, unlike our own, the two traditions could not easily be "divided and distinguished." Both were the products of man's thought, experience, and imagination, and man, for Cooper, as for his contemporary Sir Thomas Browne, is

> that great and true *Amphibium*, whose nature is disposed to live not onely like other creatures in divers elements, but in divided and distinguished worlds; for though there bee but one world to sense, there are two to reason; the one visible, the other invisible. (*Religio Medici*, Part I, Section 34)

I would like to thank the staffs of the Bodleian Library, British Library, Wellcome Institute for the History of Medicine, Dr. Williams's Library, Henry E. Huntington Library, and the William Andrews Clark Memorial Library for their many kindnesses and help in providing the works and information needed to prepare this edition. I am grateful as well to Dr. M. Foot and the Bibliographical Society for permission to quote from Plomer's *Dictionary of the Printers and Booksellers* in my Introduction and to the Bodleian Library for permission to reproduce the illustrations from Cooper's works.

The preparation and production of this book have been assisted by many whose help and advice I wish to acknowledge: at an early stage, Allen G. Debus offered suggestions as to content and form; Michael J. B. Allen kindly read a late draft of the Introduction; Nicolas Kiessling shared his bibliographical knowledge, and I received useful information from G. Groom, Leonard Beck, Peter Krivatsy, Phil Lapsansky, and Jeffrey L. Sturchio. The efforts and expert skills of Thomas C. Faulkner, the assistance of Rhonda L. Blair, and the facilities of the Humanities Research Center at Washington State University proved invaluable in providing computerized typesetting and the preparation of camera-ready copy. I would also like to thank Mr. Dale Hearell for his able assistance with library research, typing, and proofreading. Support for this project was provided in part by funds from the National Endowment for the Humanities and by the Washington State University Graduate School. Finally, my wife, Lucy, offered help and encouragement from the time the project began: in the Duke Humfrey Library at the Bodleian during a cold Oxford winter.

Introduction

William Cooper and the *Catalogue of Chymicall Books:* Science and Occultism in Restoration England

William Cooper, the late seventeenth-century author, editor, publisher, bookseller and book auctioneer who conducted his varied enterprises at the sign of the Pelican in Little Britain, London, lives today in much the same obscurity among humanists and historians of science, medicine, and the occult, that seems to have marked his existence among his contemporaries. His name does not appear in the *Dictionary of National Biography* or the *Dictionary of Scientific Biography*; it is generally absent from histories of science, chemistry, and alchemy and has, surprisingly, even escaped inclusion in Lynn Thorndike's monumentally inclusive *History of Magic and Experimental Science*. Despite this neglect, Cooper is an interesting and important figure and, particularly as compiler of the first catalog of chemical, alchemical, medical, and hermetic books in English, one whose work is valuable to the student of early modern scientific and intellectual history.

Echoing the few brief comments that have been devoted to Cooper since the beginning of this century, a recent bibliographer has observed that Cooper "is among the most remarkable of English publishers and booksellers and his relative obscurity is undeserved."[1] Except for his career as England's first book auctioneer, which has received detailed study by Lawler,[2] information about his life is extremely scant, conjectural, and, at times, contradictory; nonetheless, on the basis of such evidence, accounts like the following from Plomer's *Dictionary of the Printers and Booksellers . . . from 1668 to 1725* have appeared:

> COOPER (William), bookseller and book-auctioneer in London, (I) Warwick Lane; (2) The Pelican, Little Britain, 1668–88. In a volume of astrological tracts called *The Philosophical Epitaph*

---

1 Ian Macphail, ed., *Alchemy and the Occult: A Catalogue of Books and Manuscripts from the Collection of Paul and Mary Mellon Given to Yale University Library* (New Haven: Yale University Press, 1968), II, 449. Other brief references to Cooper's life and career are included in John Ferguson, *Bibliotheca Chemica* (1906; rpt. London: Verschoyle, 1954), I, 135, and "Some English Alchemical Books," *Journal of the Alchemical Society*, 2 (1913), 10–11; Ronald S. Wilkinson, "The Problem of the Identity of Eirenaeus Philalethes," *Ambix*, 12 (February, 1964), 25–43, and other passing references in articles by Wilkinson cited in Appendix II, notes 3 and 4.

2 John Lawler, *Book Auctions in England in the Seventeenth Century* (1676–1700) (1898; rpt. Detroit: Gale, 1968), pp. 1–45.

> *of W. C., Esquire*, 1673, it appears that at an early period of his life William Cooper was a religious enthusiast and a great believer in astrology, and that he either suffered from some physical illness, or that his religious beliefs got him into trouble with the authorities. In the dedication to the *Epitaph* he speaks of having been "in a living grave" in 1652, and having been "helped from his long troubles" by Elias Ashmole in 1662. His first entry in the Stationers' Register was made on January lst 1667 [i.e. 1667/8] and was an astrological tract called *A Chymicall dispensatorie, or, the Treasurie of medicines, written in Latine by John Schroder and Englished by William Rowland.* [*Stat. Reg., Roxb.* II. 383.] To his *Epitaph* he added "A Catalogue of Chemical Books" to which in 1675 he made several additions; this was the first English bibliography of the subject. Cooper was also the first book-auctioneer in England. His first auction was held in 1676, the library dispersed being that of Dr. Seaman. From that time until his death he shared with Edward Millington the foremost place as an auctioneer, amongst the more important libraries that came under his hammer being those of Dr. Samuel Brooke, 1680, Dr. Nathan Paget, M.D., 1681, Walter Rea, 1682, and the stock of Richard Davis, bookseller of Oxford, between 1686 and 1688. Amongst Cooper's publications may be noted Spenser's *View of the state of Ireland,* Warc's *Dc Scriptoribus Hiberniae*, Usher's *Sylloge veterum Epistolarum Hibernicarum*, an edition of Head's *English Rogue*, and Marlowe's *Lust's Dominion.* Cooper died before Nov. 1689, being succeeded by his widow.[3]

Plomer's account either errs or is misleading on several points. He persists, for example, in confusing astrology and alchemy, and his conjecture that Cooper appears to have been a "religious enthusiast and a great believer in astrology" is without any real evidence. Further, Plomer attributes, without question, the *Philosophical Epitaph* to Cooper even though definitive evidence of this authorship is lacking, and, as we shall see, some contrary evidence exists. Finally, Plomer incorrectly assigns to Cooper, publication of the cited works by Spenser, Ware, Usher, Head, and Marlowe; these were, in fact, all produced by other printers and are related to Cooper only in that all were included in lists of books that Cooper published to advertise his wares at the bookshop in Little Britain (see Appendix I).

Despite such errors and omissions, not all of which originate with Plomer, this remains the fullest general account of Cooper's life and career. It provides an outline to be corrected and fleshed out with addi-

---

[3] Henry R. Plomer, *A Dictionary of the Printers and Booksellers who were at work in England, Scotland and Ireland from 1668 to 1725* (1922; rpt. Oxford: Bibliographical Society, 1968), pp. 80–81.

tional detail drawn, in the absence of biographical records, from various book catalogs and from the prefaces and dedications to the works which he either printed or had printed by others for sale at the Pelican.

Cooper's Career as Publisher and Bookseller

Extending from Long Lane and the Smithfield Markets on the north, southeastward past St. Bartholomew's Hospital, and then turning eastward to intersect Aldersgate Street at St. Botolf Church, Little Britain (along with St. Paul's Church-yard, Paternoster Row and Moorfields) was known chiefly as a bookselling district in the seventeenth century.[4] Here, for example, until 1666, was located the shop and thriving publishing business of Richard Chiswell, known primarily for his printing of sermons and other religious works (after the Great Fire, Chiswell moved to St. Paul's Church-yard).[5] Edward Millington, later famous as a book auctioneer who collaborated with Cooper, operated a shop at the sign of the Bible in Little Britain and shared his lodgings there with Milton about 1662. And it was from a bookseller on this street that the Earl of Dorset supposedly obtained a copy of the "unknown" *Paradise Lost,* showed it to Dryden, and thus helped to initiate the poem's fame.[6] The home of Sir Thomas Bodley, founder of the Oxford library, was located on Little Britain, and following the Fire and in the next century when the center of the book selling trade had shifted to the Fleet, Little Britain was the home of Samuel Buckley, publisher of the *Daily Courant* and *Spectator*, and even for a time of Benjamin Franklin.[7]

In the absence of other documents, Cooper's activities at the sign of the Pelican in Little Britain are known to us primarily through entries for the books he proposed to publish as recorded in the *Stationers' Register* or that he actually published as recorded in Donald Wing's *Short-Title*

---

[4] For a brief historical sketch of Little Britain as a publishers' and booksellers' district, see Frank Morley, *Literary Britain* (New York: Harper, 1980), pp. 21–7; George H. Cunningham, *London: Being a Comprehensive Survey* (London: J. M. Dent, 1927), pp. 416–17, and Sir Walter Besant, *London City* (London: Adam Black, 1910), p. 352 and maps.

[5] See the *OED* and Henry R. Plomer, *A Dictionary of the Booksellers and Printers who were at Work in England, Scotland and Ireland from 1641 to 1667* (1907; rpt. Oxford: Bibliographical Society, 1968), pp. 45–6.

[6] On Millington, see Plomer, *Dictionary* . . . 1668–1725, p. 207, and Cunningham, *London*, pp. 416–17. For the relationship with Milton, see Morley, *Literary Britain*, pp. 26–7.

[7] Morley, *Literary Britain*, p. 27; Cunningham, *London*, p. 417.

Catalogue, through catalogs that he himself compiled to advertise his book auctions, and through the brief and incidental comments contained in the prefaces, dedications, advertisements, and notices of other kinds that are included in the works he authored, published, edited, translated, and sold. Though providing nothing like a full biographical sketch, these sources at least shed light on his activities and interests during a period of over twenty years.

While it must be remembered that entries in the *Stationers' Register* are records of proposed books and not of those which were necessarily published, they serve nonetheless in establishing approximate dates for Cooper's publishing activity and indicate as well his interests and specializations. Entries assigned to Cooper exist for a period of twenty-one years, from 1 January 1667 to 2 December 1687, and include records of seventeen proposed titles. Of these, twelve resulted in books directly associated with Cooper, either through having been printed by him or by others for sale at his bookshop.[8] As we shall see shortly, Cooper published a number of other titles which were not entered in the *Stationers' Register*; however, among the twelve duly entered we find strong preferences for books dealing with alchemy and chemical medicine and especially those authored by the mysterious Eirenaeus Philalethes (see Appendix II). The first of these volumes, *Secret's Reveal'd: or, an Open Entrance to the Shut Palace of the King,* was entered on 15 September 1668 and printed by W. Godbid for Cooper "in Little St. Bartholomews, near Little Britain" in 1669.[9] Other Philalethean tracts include the Latin *Opus Tripartitum de Philosophorum Arcanis* and *Ripley Reviv'd:*

---

8 As Plomer notes, another volume, J. Schroeder's *Chymical Dispensatory* (entry no. 333), entered on 1 January 1667, was the first entry assigned to Cooper in the *Stationers' Register*; however, it was actually published by John Darby in 1669 and not by Cooper.

9 *The Epistle Dedicatory* of this volume, "To the Right Honourable Lord Lucas, Baron of Shenfield in Essex," dated 15 September 1668, is of interest for its autobiographical information as well as its anticipation of later dedications to the great and famous, i.e., Ashmole and Boyle, in the *Philosophical Epitaph*. Cooper admits that he is not personally acquainted with Lord Lucas, "though living formerly where your Lordship did frequent, and being a Neighbor in our Youthful Years" (sigs. A3, A3^{v}). The Dedication has been prompted by Lucas' "Eminent Discourses of deep Mysteries" with a friend of Cooper's, "a most Learned Frenchman, . . . at your return from Oxford, in the late Rebellious Times." Having thus been attracted by Lucas' apparent interest in some form of hermeticism, Cooper was eager to acquaint him with the present work by Eirenaeus Philalethes ("this English rare Phoenix of Learning"). He concludes, as in the Ashmole dedication of the *Philosophical Epitaph*, with vague reference to a time of "long Troubles." The dedicatee was John Lucas, born 23 October 1606, the son of Sir Thomas Lucas of St. John's near Colchester. He was a strong supporter of Royalist causes and "was imprisoned, but escaped, and fought at Newbury." See *The Complete Peerage*, by G. E. C., rev. Vicary Gibbs, ed. H. A. Doubleday and Howard de Walden (London: St. Catherine Press, 1932), 8, 245.

or, an Exposition upon Sir George Ripley's Hermetico-Poetical Works, both entered on 30 February 1678 and published in the same year. (Curiously, the *Opus Tripartitum* was entered a second time on 29 August 1683, again to Cooper, but there is no evidence that this collection was republished.)

In terms of *Stationers' Register* entries for alchemical books and those dealing with chemical medicine—in fact, for books of all sorts—1683 was Cooper's most prolific year. On 27 August the ten essays which comprise the extremely important *Collectanea Chymica* (printed for Cooper in 1684) were entered, including one by Eirenaeus Philalethes, one by George Starkey, with whom Philalethes is often identified or closely related, and others by such revered alchemical authorities as Roger Bacon, Sir George Ripley, the Earl of Trevisan, and Jean Baptiste van Helmont. One day later, on 28 August, Cooper entered a second important collection, the *Aurifontina Chymica, or, a Collection of Fourteen small Treatises Concerning the First Matter of Philosophers*. This "curious and very rare little volume," as Ferguson describes it,[10] which Cooper appears to have compiled and edited himself, and which had been printed for him in 1680, included a dedication of its opening treatise to Charles II,[11] and essays by Ripley, Nicholas Flamel, Raymond Lull, the Earl of Trevisan and several *anonymi*. On 29 August, two additional entries appear for occult books: the first is for Sir Kenelm Digby's *Chymicall secretts, and rare experiments in physick and phylosophy . . . published by George Hartman his steward*, also previously printed for sale by Cooper in 1682 and reissued. The other is the aforementioned "second" entry for the *Opus tripartitum*, of which there is no actual republication evidence.

The lengthy series of entries appearing in August, 1683, concludes with three dated on the 30th. The first, containing puzzling bibliographical problems to which I shall turn in a later section, is for *A phylosophicall epitaph in hieroglificall figures . . .* to which is added *A cattalogue of chymicall bookes collected by Wm Cooper, bookseller*. There follows an entry for a work by John Dowell entitled, *The Clergies hour, or, the lifes of St. Bassill the great, and St. Chrysostum*, which seems never to have been published. Finally, there is an entry for Thomas Houghton's popular book on English miners and mining, *Rara Avis in Terris; or, the Compleat Miner*, a work that had actually been published two years earlier in 1681.

There is now a hiatus of about two and a half years in *Register*

---

10 Ferguson, I, 58.

11 This volume is usually listed under the name of John Frederick Houpreght, who wrote the *Epistle Dedicatory* to Charles II and also translated the first treatise in the volume, the *Hydropyrographum Hermeticum*.

entries, until 26 March 1686 when the *Ephemeris ad ann a Nativitate Domini 1686* (Wing A1805) was published. The final listing, occurring on 2 December 1687, is for printing William Barton's *Six centuries of select hymns and spirituall songs*, which was published as a fourth edition in Barton's popular series in 1688.

Thus, on the basis of the evidence provided by entries in the *Stationers' Register* and from the books which were actually published, Cooper's preferences are clearly for books dealing with alchemy, chemistry, medicine, the occult, and the difficult-to-distinguish blendings of these fields: those shadowy borderlines that characterize much seventeenth-century science and pseudo-science. Of the seventeen entries assigned to Cooper, only four, *The Wonders of the World . . . collected by Rich: Smith* (29 March 1671) and apparently not published, *A French Grammer* [sic] (4 December 1673), and the previously cited works by Dowell and Barton, fall clearly outside these categories. A second tendency, equally pronounced, is the irregularity of the distribution of entries over this twenty-one year period, with long intervals occurring between several of them, e.g. 1668–71–73–75–78–83, and the heavy concentration of entries—seven of the seventeen—recorded in 1683.

Cooper's interest in the production and sale of occult and scientific books is further confirmed by more detailed consideration of other works, i.e., those not entered in the *Stationers' Register*, that he either published himself or had published by others for sale at the sign of the Pelican. As reflected in the entries of Wing's *Short-Title Catalogue*, Cooper's activity as publisher and bookseller extends from 1669, the year of the aforementioned *Secrets Reveal'd* (Wing S5288) until 1689. In 1670, he was one of two designated London booksellers for William Barton's *Two Centuries of Select Hymns and Spiritual Songs*, printed by W. Godbid; in 1672, another edition of Barton's work appeared. In 1673, the *Philosophical Epitaph*, containing the first, shorter version of the *Catalogue of Chymical Books*, was printed for Cooper; a three-part version of the latter, which is the basis for the present edition, was produced in 1675 and was extended in the *Continuation or Appendix* of 1688.

In addition to the expanded version of the *Catalogue*, 1675 also saw publication, for sale by Cooper, of two of William Simpson's contributions to the common seventeenth-century debates concerning the efficacy of spas and mineral water: the *Zymologia Physica, or a brief Philosophical Discourse of Fermentation* appeared along with *A Discourse of the Sulphur-Bath at Knaresbrough in Yorkshire* (entry no. 350). As he frequently did (see Appendix I), Cooper used the opportunity provided by the publication of Simpson's popular works to advertise other books that he knew would appeal to his readers. Following the final page of the text of Simpson's *Discourse*, he added a two-page list of "Certain Books Printed, and to be Sold at the Pellican in Little-Britain, London." Not all of the titles included are those of chemical books, but at the end of the list, and in reference to his bookshop, he adds the note: *Where likewise most sorts of Chymical or Philosophical Books, may be had, whether in English or Latin.* In a second advertisement placed between the two parts of the Simpson volume, Cooper makes the timely announcement of the publication of the expanded *Catalogue of Chymicall Books*, which appeared in the same year.

Cooper's fourth publication of 1675, the funeral sermon for Dr. Lazarus Seaman, "Late Pastor of Allhallows-Bread-street, London," titled "Exodus: or, The Decease of Holy Men and Ministers" and preached by William Jenkyn, "late Minister of the Gospel at Christ-Church, London," serves to link Cooper's career as a publisher with the beginning of his career as England's first book auctioneer. Seaman's library was the first that Cooper sold at auction, on 31 October 1676, and its contents are listed, along with much interesting information on his auctioneering procedures, in the *Catalogus Variorum & Insignium Librorum . . . Viri Lazari Seaman*, which he published in 1676 (Wing S2173).

During the next two years, Cooper's publishing activities were devoted exclusively to works that were authored under the closely-related names of Eirenaeus Philalethes and George Starkey.[12] In addition to the Philalethean works previously cited, Wing includes separate entries for four of the individual works that Cooper included in *Ripley Reviv'd* (entry no. 295): *An Exposition upon Ripley's Epistle to King Edward IV, An Exposition upon Sir George Ripley's Preface, A Breviary of Alchemy*, and *An Exposition upon Ripley's Vision.* However, Cooper's interest in hermetic publications was not restricted to works written in English or to English translations, as we see from the *Opus Tripartitum de Philosophorum Arcanis* of 1678. This work, in Latin, contains three Philalethean tracts: the *Enarratio methodica trium Gebri medicinarum*, the *Experimenta de praeparatione Mercurii Sophici*, and the *Vade mecum philosophicum.* Wilkinson reports that Cooper also published an English translation of the *Opus Tripartitum* later in the same year,[13] but I have been unable to confirm the existence of this edition. Cooper's continuing interest in Philalethean works is further seen in the list of twenty-eight items, including manuscripts, which is contained in the "*Catalogue* [of books] written by *Eirenaeus Philalethes*" printed as an advertisement and published with the *Exposition upon Sir George*

---

12 On the problems of identifying these two authors and their relationship, see Appendix II and especially Ronald S. Wilkinson, "The Problem of the Identity of Eirenaeus Philalethes," *Ambix* 12 (February, 1964), 25–43.

13 "Problem of the Identity," 32.

Ripley's Epistle (1677) and *Ripley Reviv'd.* (For a transcription of this *Catalogue*, see Appendix II.) An entry for a final Philalethean treatise, *The Secret of the immortal liquor called alkahest* is dated 1683, although the work was actually published in the *Collectanea Chymica* of the following year. Thus, it is clear that Cooper's role in the dissemination of Philalethean materials was extremely critical, and A. E. Waite has called this anonymous author, "the most important expositor of Alchemy in the second half of the seventeenth century."[14]

While chemical, medical, and hermetic books remained Cooper's specialty to the end of his publishing career, his interest in other types of scientific literature is also apparent in the Wing entries. An edition of Gabriel Plattes' popular *Discovery of Subterraneall Treasure*, originally published in 1639, was printed for sale at the Pelican in 1684, and thus continued to serve as a means of encouraging experiment and innovation in the fields of mining and metallurgy. Cooper's interest in English mining is further demonstrated by the publication of four editions of works by Thomas Houghton: the *Rara Avis in terris; or, the Compleat Miner* of 1681 (entry no. 208); *The Laws . . . of the Miners in the Forrest of Dean* and the *Ancient Laws . . . of the Miners in the King's Forrest of Mendipp*, both of 1687 and combined by Cooper in the following year to form two of the three parts of a new edition of the *Compleat Miner* (entry no. 207).

In addition to such technological and scientific works, Cooper's record during his last few years as publisher and bookseller includes titles of surprising diversity. In 1685, for example, Raymond Minderer's *Minerva: The High-Dutch Grammer* [sic] was printed for him, and three impressions of Edmond Halley's *Ephemeris ad annum* appeared in 1686–88. Also in 1688, the penultimate year for which Wing entries are recorded, Cooper returned to an author and type of work with which his publishing career had begun: William Barton's *Six Centuries of Select Hymns*, 4th edition, was produced for sale at the Pelican.

During Cooper's final years as a publisher, however, works on such non-occult subjects are clearly exceptional, and, as in the earlier years, the majority of his titles reflect his abiding interests in alchemy and chemical medicine. The period from 1685–89 shows a particular preference for collections of such treatises, ancient and modern, foreign and English. Two posthumous works of Sir Kenelm Digby, the *Chymical Secrets. Second Part* and his *Choice Collection of Rare secrets* (entry no. 120), had been printed for sale by Cooper in 1682; followed, in 1686, by *The Works of Geber* (Jābir Ihn Haiyān, entry no. 212) in the translation of Richard Russel, and, in 1688, by Francis Mercurius van Helmont's

---

14 *The Secret Tradition in Alchemy* (London, 1926), p. 206, as quoted in Wilkinson, "Problem of the Identity," 24.

One Hundred Fifty Three Chymical Aphorisms (entry no. 183) in the translation of Christopher Packe. Finally, in 1689, a lavishly printed and illustrated folio edition of *The Works of the Highly Experienced and Famous Chymist, John Rudolph Glauber*, again translated by Christopher Packe, was printed by Thomas Milbourn and authorized for sale in the bookshops of the printer, of Cooper, and of D. Newman at the King's Arms in the Poultry. This volume reveals another way in which Cooper assisted the production of important and expensive chemical and alchemical books: he subsidized the translation and publication costs as a subscriber to the project. "Mr. *William Cooper*, Bookseller" appears in "A CATALOGUE of the Subscribers Names" which is included at the beginning of the Glauber volume. However, the actual terms of subscription appear, not in this volume, but in F. M. van Helmont's *One Hundred Fifty Three Chymical Aphorisms*; these terms set forth precisely the obligations of the subscribers and provide an interesting glance at such publishing arrangements in the seventeenth century:

> That whosoever shall pay to *Christopher Pack*, at the sign of the *Globe* in the Postern-street, near *Moorgate*; or to *William Cooper*, at the *Pelican* in *Little-Britain*, Bookseller; or to *Dorman Newman* at the *Kings-Arms* in the *Poultry*, Bookseller; the sum of twenty shillings, in the manner following, shall have an Compleat Book in Quires, viz Ten shillings in hand, and Ten shillings more upon the delivery of the said book, which is intended to be published in *Trinity*-term next. And for Encouragement of those who shall subscribe for six Books, they shall have one Book, when printed presented them *Gratis*. The *Specimen* may be had at any of the three places above-mentioned, as also of divers Booksellers in and about London.[15]

Milbourne's "Post-Script to the Reader," in addition to describing the volume's contents, concludes with the following plea:

> These are the Treatises of which I could not give a particular Account in the Specimen, and which being added to those there exhibited, make in all 55 Treatises, which will amount to a large Volume in Folio, in which are also many Cutts [i.e., engraved illustrations]; which is all ready for the Press, and wants nothing but a sufficient number of Subscribers to give it Breath in the English Air; I have spared neither Paines nor Cost to serve my Country in this Work; and I hope there will not be wanting so many Generous Souls, who will concur with me (by subscribing to it) to render it

---

15 Franciscus Mercurius van Helmont, *One Hundred Fifty Three Chymical Aphorisms* (London, 1688), p. 63.

effectual.[16]

Although by subsidizing publication of Glauber's *Works* Cooper helped assure his own future sales and profits, it is quite likely that he would have agreed with Milbourne that by doing so he was also "serv[ing his] Country in this Work." Financial support of significant publication projects, though perhaps less important than his actual printing and selling of books, was vital to the enterprise and to the dissemination of chemical, occult, and medical knowledge, in particular. To these ends, Cooper also added his talents as an auctioneer and author.

Cooper and Book Auctioneering in England

Primarily as the result of a large number of extant sale catalogs, our knowledge of Cooper's activities as England's first book auctioneer is more complete than any other aspect of his career. Drawing chiefly on these as a source, John Lawler, in *Book Auctions in England in the Seventeenth Century* (1676–1700), has assembled much information on the inclusive dates of Cooper's auctioneering career, the frequency and dates of many of his sales, his clients, and, most important, estimates of the number of books included in certain of the sales and listings of their more notable contents. Lawler reports, for example, that Cooper's auctioneering began on 31 October 1676 with the sale of the Seaman library, included at least twenty-two sales over a thirteen year period, and ended in June, 1688.[17] While Lawler's research provides us with much quantitative data which illuminates the flourishing book auction business in Restoration England, it adds little to our understanding and knowledge of Cooper himself. And in several instances, Lawler is inaccurate in his treatment of purely factual matters. For example, after stating (correctly) in his Introduction that occult books were Cooper's principal publishing interest, Lawler reports that "In 1669 he [Cooper]

---

16 Helmont, p. 62.

17 Lawler, chapter two and, especially, pp. 215–24. Neither Lawler nor any other scholar has noted that the auctioneering fame of Cooper and his partner Millington was celebrated in George Smalridge's Latin poem, *Auctio Davisiana Oxonii habita, per Gul. Cooper & Edward Millington Bibliopol.*, published in London, 1689. According to Anthony Wood, Smalridge was born in Lichfield and matriculated at Christ Church, Oxford, in 1682. Wood reports that "this excellent Latin poem was written on the sale of the books of Rich. Davis, an ancient bookseller of Oxon, which were exposed to sale by way of auction in a large stone fabric, opposite to St. Michael's Church in Oxon, near the north gate of the city, commonly called Bocardo," *Athenae Oxonienses*, ed. Philip Bliss (London: Rivington, 1813–20), 4, 667.

had printed for him by Mr. Godbid in St. Bartholomew's Close, a 'Treatise of Richard Vaughan the Alchemist' (Eirenaeus Philalethes) under the following quaint title: *Secrets Revealed, or an Open Entrance to the Shut Pallace of the King.*"[18] Here Lawler has, of course, confused Eirenaeus Philalethes, whose identity is uncertain but who is often associated with George Starkey, and Eugenius Philalethes, the pseudonym of Thomas Vaughan, the well-known hermetic writer and brother of the poet Henry Vaughan. Immediately thereafter, he seriously misquotes the title of *The Philosophical Epitaph of W. C. Esquire*, to which I will turn shortly, calling it *The Philosopher's Epitaph.*[19]

Despite such errors, Lawler can not be faulted for having found little to report about Cooper's life. For one who actively engaged in a variety of enterprises centered in the book trade—as printer, author, editor, translator, bookseller—and for a period of over twenty years, it is surprising that details of his life, public and private, are almost non-existent. Although Lawler generally avoids subjective interpretations, his analysis of the sale catalogs which Cooper used to advertise his auctions suggests that he was innovative, shrewd, and industrious. From Lawler's account, which need not be reiterated in detail, we learn that Cooper probably patterned his methods and practices after those of the Elzevir family in Leyden and Amsterdam, which had been auctioning books since the early years of the seventeenth century.[20] That Cooper was aware of his historic role in this new method of bookselling in England, that he recognized its importance for the transmission of knowledge, and that he understood the need to establish a format and procedures to regulate the business (it is not difficult to regard him as the typical "Puritan" in business matters) is clear from the address "To the Reader" that prefaces the Seaman catalog:

> It hath not been usual here in *England* to make Sale of Books by way of *Auction*, or *who will give most for them*: But it having been practised in other Countreys to the Advantage both of Buyers and Sellers; It was therefore conceived (for the Encouragement of Learning) to publish the Sale of these Books this manner of way; and it is hoped that this will not be unacceptable to Schollers; and therefore we thought it convenient to give an Advertisement concerning the manner of proceeding therein.
>
> *First*, That having this Catalogue of the Books, and their Editions under their several Heads and Numbers, it will be more easie for any Person of Quality, Gentlemen, or others, to Depute any one

18 Lawler, p. xxvi.

19 Lawler, p. xxvii.

20 Lawler, p. 1.

> to Buy such Books for them as they shall desire, if their occasions will not permit them to be present at the *Auction* themselves.
>
> *Secondly*, That those which bid most are the Buyers; and if any manifest Differences should arise, that then the same Book or Books shall be forthwith exposed again to Sale, and the highest bidder to have the same.
>
> *Thirdly*, That all the Books according to the Catalogue are (for so much as we know) perfect, and sold as such; But if any of them appear to be otherwise before they be taken away, the buyer shall have his choice of taking or leaving the same.
>
> *Fourthly*, That the Money for the Books bought, be paid at the delivery of them, within one Months time after the Auction is ended.
>
> *Fifthly*, That the *Auction* will begin the 31st of October at the Deceased Dr's. House in *Warwick Court* in *Warwick Lane* punctually at Nine of the Clock in the Morning, and Two in the Afternoon, and this to continue daily until all the Books be Sold; Wherefore it is desired, that the Gentlemen, or those Deputed by them, may be there precisely at the Hours appointed, lest they should miss the opportunity of Buying those Books, which either themselves or their Friends desire.[21]

With only slight variations, these procedures are repeated in nearly all the catalogs which Cooper printed throughout his career.

Similarly, there is a high degree of consistency in the formats of the catalogs themselves, the categories varying only slightly according to the types of books which the libraries contained. Each entry generally receives one closely printed line and includes the author's last name (occasionally the first), a short title, the place and year of publication, and sometimes the size of the volume, although the categories themselves often depend on book size as well as subject. Entries are generally numbered but not arranged alphabetically by author.

Large sales, such as that of the Seaman library or the three-part sale to liquidate the enormous stock of the Oxford bookseller, Richard Davis, required numerous and complex categories and sub-categories, such as Patres Graeci and Latini in folio and quarto, Rabbini, Philology Graeci, Libri Scholastici, Tractatus Theologici, Concilia, Scriptores Ecclesiastice, as well as Philosophy and Mathematics, Medicine, English Physick, Books in French, and English Miscellany in folios, quartos, octavos and 12's. In contrast, catalogs for smaller and more homogeneous libraries are arranged with few specialized subject classifications or may

---

21 *Catalogus Variorum & Insignium Librorum Instructissimae Bibliothecae Clarissimi Doctissimiq; Viri Lazari Seaman, S. T. D. . . . Cura Gulielmi Cooper Bibliopolae.* Londini Apud Ed. Brewster & Guil. Cooper, 1676, "To the Reader," immediately following t. p.

depend entirely on classification by volume size.[22]

While the information provided by Cooper's sale catalogs is overwhelmingly factual and statistical, on rare occasions his prefaces provide unexpected insights into his life and times. In the catalog which advertises the sale of the library of Dr. William Outram (12 December 1681), for example, Cooper writes that in addition to the holdings of this Nuper Ecclesiae Westmonasteriensis Canonici, "so much of the library of Mr. Thomas Gataker, as was rescued out of the Flames of London" would also be included in the auction.[23] This statement makes one aware of another of the disastrous consequences of the Great Fire and recalls Sir Walter Besant's observation, that "some of the people placed their things for safety in the churches, fondly thinking that the fire would spare them; the booksellers of Little Britain and Paternoster Row deposited the whole of their books in the crypt of St. Paul's; alas! they lost them all."[24] Although we do not know precisely when Cooper opened his shop, it is entirely possible that his business was disrupted by the fire.

The generally impersonal style of these prefaces is also briefly interrupted in the auction catalog for the library of Rev. Daniel Rogers, which was sold beginning 21 June 1683; here Cooper's efforts to publicize the diverse contents of this library contain a brief glance at the English Civil War:

> Many parts of the Catalogue might be instanced in to recommend the Books and their buying, to all unprejudiced Persons. viz. That great variety of choice Books in Divinity, History, Philology, in all Volumes, and of most Subjects: As also, those curious Volumes of Tracts, and that exact Collection of Forreign and Domestick News [From 1640 to 1682], of all sorts that was published in the beginning, and during our unhappy Civil-Confusions with all others of that Nature since our happy Settlement, methodically digested and continued to these times, is enough to intimate by way of Invitation to all Lovers of Learning, and Knowledge.[25]

While Cooper does not, as a rule, use his prefaces as means of

---

22 See, for example, "An Appendix of Books to be sold by Auction on 2 June 1679," i.e., the libraries of Stephen Watkins and Thomas Sherley; also Lawler, p. 15.

23 From the *Sale of the Library of Dr. William Outram*, title page, verso.

24 Sir Walter Besant, *London in the Time of the Stuarts* (London: Adam Black, 1903), p. 245. A more detailed and graphic account of this incident, which is the source of Besant's description, is in Lord Clarendon's *Life*; quoted in Edward Arber, *Term Catalogues* (London, 1903), I, vii–viii.

25 *Catalogus Variorum Librorum* (Bodleian MS. Ashmole 1058), "To the Reader," leaf 2, recto.

exerting undue pressure on his readers to attend auctions and purchase books, he will, as in the above instance, occasionally use them to commend certain types of books to the attention of his audience. In particular, he is anxious to inform readers of sales containing important collections of alchemical, chemical, and medical books as well as those that fall into the general category of natural philosophy; thus his interests as an auctioneer coincide closely with his interests as publisher and bookseller.

Given these preferences, Cooper must have been especially enthusiastic about the sale of the library of Nathan Paget, the medical doctor and friend of John Milton, which took place at the Pelican on 24 October 1681.[26] While this library contained very substantial holdings of Latin works in theology and philology (605 lots), and a great quantity of English books on all subjects and in all sizes, its major concentration (816 lots) was Latin works on medicine, chemistry, mathematics, and natural history. The *Bibliotheca Medica Viri Clarissimi Nathanis Paget, M.D.,* which Cooper assembled to advertise this sale, contains 52 pages, comprising, according to Lawler, approximately 2600 lots and 7000 to 8000 titles.[27] Lawler gives special attention to the occult books contained in Paget's library:

> The most interesting feature about the collection is the very large number of books coming under the heading of the "Occult Sciences," . . . Here we find the early writers on such abstruse subjects as Astrology, the Philosopher's Stone, the Elixir of Life, Ghosts, Apparitions, Witchcraft, Possession by Demons, Exorcisms, and other speculations on the unknowable, . . . the writings of Cardan, Paracelsus, H. Cornelius Agrippa, Roger Bacon, Robert Fludd, Van Helmont, Jacob Behmen, Wm. Lilly, Raymund Lully, J. B. Porta, Basil Valentine, J. R. Glauber and many lesser known and anonymous writers dealing with similar subjects.[28]

---

26 The *Bibliotheca Medica Viri Clarissimi Nathanis Paget, M.D. . . . Per gulielmum Cooper Bibliopolam.* includes the following entries for works by Milton: "English Books in Quarto": 94 Joh. Miltons Eikonoclastes or answer to K. Charles his Eikon Basilikae; 179 Jo. Miltons Paradise Lost, a poem in ten books; 180 Another of the same; 217 Jo. Miltons Apology for Smectymnus with the reason of Church-Government; 249 John Miltons Areopagitica or speech to the Parliament for liberty of unlicensed Printing; 294 Jo. Miltons Tetrachordon expounding 4 places of Scripture for Nullities in Marriage 307 Jo. Miltons exposition upon the 4 chief places in Scripture which treat of Marriage, &c. "English Books in Octavo": 319 Jo. Miltons Paradise regain'd in 4 books; 363 Jo. Miltons Paradise lost a poem in 12 books; 510 Jo. Miltons Treatise of Civil Power in Ecclesiastical Causes against Force in Religion.

27 Lawler, p. 19.

28 Lawler, p. 19.

Further evidence of Cooper's special interest in works of this type appears in the catalog describing the contents of the libraries of the Reverend Richard Chace and two medical doctors, Christopher Bathurst and George Tonstall (see entry no. 384), that were sold in Little Britain on 24 March 1683–4. Cooper notes in his preface that

> It is not Customary with me to make long Harangues to set off the goodness of our Books, (I have no reason to question the Judgements of the Buyers,) but yet I thought it not impertinent at this time to acquaint the Reader, that here is a very good Collection of Chymicall as well as Medicinall, and likewise of Philologicall as well as Theologicall Books both in Latin and English.[29]

Again in the catalog for the sale of the libraries of the Reverend Dr. Stokes and others, which began on 1 December 1685, Cooper singles out the strengths of the collection, here including the many chemical and alchemical books in the "Mathematics" and "Physica" categories.[30] Finally, a rather small but important sale of a medical collection occurred on 13 July 1686 and included a total of 487 lots, many of which contain titles on chemical medicine. As in the Chace catalog, Cooper uses the address to the reader to advertise the excellence of his wares:

> To recommend this Collection of Physick Books I conceive needless, they are sufficient of themselves to encourage the Buyers; but yet I must affirm that the like hath not hitherto been exposed by this way of sale; for although the number of Books contained in this Catalogue are not many, yet the excellency of them are such, as the like in Auction hath never been exposed to sale, they being a collection of the best and most modern Authours both in Physick and Chyrurgery as well as Chymestry, of the last and best Editions; wherefore not questioning your encouragement, I proceed to the Time, Place and Conditions of Sale.[31]

Cooper and *The Philosophical Epitaph*

While it is tempting to attribute Cooper's interests in occult and scientific books (and those of other types) to motives wholly mercenary—

---

29 *Catalogus Librorum Bibliothecarum* (Bodleian MS. Ashmole 1058).

30 *Catalogus Librorum Theologicorum, Philologicorum,Mathematicorum, &c.* (Bodleian MS. Ashmole 1058).

31 *Catalogus Librorum Medicorum, tam Recentiorum quam Antiquorum, e Regionibus Vicinis nuperrime Advectorum*, "To the Reader," sig. A2, in Bod. Wood E. 19.

they were commodities to be printed for sale or acquired to be sold at a profit—other factors indicate that his concern for these subjects was more deeply rooted and enduring. Chief among these is publication of *The Philosophical Epitaph of W. C. Esquire, for a Memento Mori on his Tomb-stone. . . . All Published by W. C. Esquire. With a Catalogue of Chymicall Books.* The *Epitaph* is a curiously complex work, one which presents a variety of questions and difficulties which require examination, if not definitive solution, before its appendage, the *Catalogue of Chymicall Books*, can be considered. Quite apart from questions of the meaning and interpretation of the texts of individual treatises, the illustrations, and their interrelationship, the *Epitaph* presents problems concerning authorship, dating, and the arrangement of its prefatory addresses and dedications. Let us briefly consider these concerns.

As we have seen, the few scholars and bibliographers who have taken note of the *Epitaph* have not hesitated to identify the "W C Esq" listed as its publisher on the engraved title page, with the "William Cooper att the Pellican in Litle Britain" similarly listed as the one for whom the work was printed in 1673. (See figure one.)[32] This attribution is assumed by Ferguson, Macphail, Wilkinson, and others, and the fact that the *Catalogue of Chymicall Bookes* is included on the title page immediately below "W C Esq" and that the title page of the expanded, 1675 edition of the *Catalogue* specifies "Collected by Will. Cooper, Bookseller," has strengthened the view that Cooper was responsible for the *Epitaph* as well. There is, however, one important piece of evidence which suggests that this is not the case. In a brief addendum published in his edition of the works of Elias Ashmole, C. H. Josten notes that the *Transcripts of the Stationers' Register* indicate the author as being not Cooper, but one William Chamberlaine, and that this attribution is "confirmed by the Latin motto [added under the initials W. C. at the end of the dedicatory letter to Ashmole in the *Epitaph*], '*Laurum Amice*

32 Though his name does not appear on the design, the engraver responsible for this title page is almost certain to have been Robert Vaughan, who, according to Alfred Forbes Johnson (*A Catalogue of Engraved and Etched English Title-Pages* [Oxford: Oxford U. P., 1934], pp. 55–8) worked from 1622–78. Johnson lists twenty-eight title pages for which Vaughan was responsible, including Henry Vaughan's *Olor Iscanus* (1651) and Eirenaeus Philalethes' *Ripley Reviv'd* (1678). (See figure four.) That Vaughan was particularly successful as an engraver of alchemical designs can be seen in the illustrations included in Elias Ashmole's *Theatrum Chemicum Britannicum.* My attribution of the *Philosophical Epitaph* design to Vaughan is based on its very close similarities to the signed title page of *Ripley Reviv'd*; we even see that an identical roundel ("*Corpora praeparate purgate solvite coagulate et projicite super corpus*") appears in both designs.

Eligis Rus,' which is an anagram of Guilielmus Camerarius, the Latin version of William Chamberlaine's name."[33]

Because of the complications that it introduces for both the authorship and dating of the *Epitaph*, the entry is important enough to quote in full:

30th August 1683

> Entred . . . booke or coppy entituled A phylosophicall epitaph in hieroglificall figures, with their explicacion, by Wm Chamberlaine, Esqre. A breife of the golden calfe. Glaubers golden ass well managed. Jehior with three principles, or, originall of all things, these three last published by the aforesaid W: C: Esq., to wch is added, A Cattalogue of chymicall bookes collected by Wm Cooper, bookeseller. . . . vjd
> [signed] Wm. Cooper[34]

At this stage, I can not offer a final solution to either the puzzle concerning Wm. Chamberlaine as author of the *Epitaph* or the discrepancy as to date: it actually appeared in 1673, and thus the *Stationers' Register* entry of 1683, a full decade later, must be an error on the part of the original recorder or the transcriber. As to the authorship problem, I would only add the following observations.

While it is of course possible that a William Chamberlaine who was interested in alchemy and employed cryptic forms of expression lived in the later seventeenth century, his existence is certainly not well known. The William Chamberlayne [sic] (1619–89) who is known during this period, the resident of Shaftesbury in Dorset, ardent royalist, and author of such works as the tragi-comedy *Loves Victory* (1658), *Pharonnida: A Heroick Poem* (1659), and a celebratory poem on Charles' return (1660), is a most unlikely candidate. In fact, in the "Epistle to the Reader" prefacing *Pharonnida*, Chamberlayne unleashes a withering attack on those who might easily be readers or writers of the kinds of books that Cooper dealt in; these "Popular Hermeticks," to use his term, are identified with the radical sectarians of the period and thus are banished from reading his poem:

---

33 C. H. Josten, ed., *Elias Ashmole* (1617–1692): *His Autobiographical and Historical Notes, His Correspondence, and Other Contemporary Sources Relating to his Life and Work* (Oxford: Clarendon Press, 1966), III, 1289. If "*Laurum [or L'aurum] Amice Eligis Rus*" (O Friend may you elect the country as your gold.) is derived from a literary source, I am unaware of it.

34 Edward Arber, ed., *A Transcript of the Registers of the Company of Stationers of London* (1914; rpt. Gloucester, Mass.: Peter Smith, 1967), III, 192.

> let them enter no further then the Gate; I mean, all squint-ey'd *Sectaries*, from the *Spawn of Geneva*, to the black *Brood of Amsterdam*: together with some *Rascals* of a lower Rank, such as usurp the abus'd *Title* of, *Sons of Art*, ... For the first of which, the preposterous *Genius* of the *Times* hath so far favour'd them, that now nothing is more vendible then the *surreptitious* off-springs of their immagin'd *Wit*: Every Stationers Shop affording pregnant examples of it, in big bulk'd Volumes of *Physick, Astrologie* and the like: By these indigent Vermine, either to satisfie their clamorous Wants, or inhance their esteem in the Vulgar Opinion.[35]

Until further information is unearthed on Chamberlaine or Chamberlayne and his possible connection with Cooper, it seems necessary to conclude that, for whatever reasons, Cooper used the name "Chamberlaine" as a pseudonym when entering the *Epitaph* in the *Stationers' Register*. An ancient practice in alchemical and hermetic writing, the use of pseudonyms showed no signs of diminishing in the later seventeenth century.

But more positive evidence exists to link the authorship of the *Epitaph* with William Cooper the bookseller. On the illustrated page from the *Epitaph* entitled "*An* EPITAPH *made by* W. C. CLowDeD by threatnIng DIsasters" (see figure two) and which is given over to the representation of hieroglyphical scutcheons and versified mottos, there appears a curious design which Cooper later used as a printer's device on title pages of works that he published. This design, in the upper left, features a circular owl's head with triangular features, within a square, within two larger triangles which thus form a six-pointed star. The whole is surrounded by an ivy wreath and bears the caption: "Minerva's Owl in an Ivie Bush. *Creation, Chaos, Corruption*" This symbol for "philosophical" salt appears for the first time in W. C.'s *Epitaph* of 1673; however, Cooper, having some interest in the design, retained it for later use. While it does not appear on all or even a majority of subsequent title pages, it seems to have been used, according to my incomplete census, only on title pages of alchemical works attributed to Eirenaeus Philalethes: *An Exposition Upon Sir George Ripley's Epistle* (1677) and *A Breviary of Alchemy, Ripley Reviv'd*, and the *Opus Tripartitum*, all of 1678. The use of this symbolic design strengthens the link between W. C., the author of the *Epitaph*, and William Cooper the printer, and it may also have been the printer's means of associating publication of his favorite alchemical author with the sign of the Pelican. We shall examine this illustration and related ones later in this section.

---

35 William Chamberlayne, *Pharonnida: A Heroic Poem* (London, 1659), "Epistle to the Reader."

A final detail relating to the authorship problem merits comment, though more for the complications it introduces than for the solution it promises. In the 1673 version of the *Catalogue*, the entry for the *Philosophical Epitaph* (appearing under the P's and concluding with "All published by W. C. Esq; Lond. 1673, 8." is immediately followed by "— *His Secrets of Alchimy*. London. 8." (see entry no. 106), indicating that the latter work was also authored by W. C. There is, however, no evidence that a work by this title was written by a W. C. or a William Cooper, or a William Chamberlaine or Chamberlayne, and it is virtually certain that this reference was placed erroneously beneath the initials W. C. Cooper doubtless intended that it should have been placed under "Paracelsus," who follows according to his faultily alphabetized system, because in the *Continuation* (sig. D*18) he uses *Secrets of Alchimy* as the main title for the *Supreme Mysteries of Nature, Occult Philosophy*, &c. (entry no. 285). Nonetheless, because of the implications that this entry may have for the identity of W. C. and the question of the authorship of the *Philosophical Epitaph*, I have chosen to retain it in its original position.

Since most of the copies of the *Philosophical Epitaph* which I have consulted have either sections missing or vary greatly in the ordering of their constituent parts, the overall arrangement of the work is at best confusing. Bodleian 8° E. 63. (2) Art., the most complete version I have consulted, includes the following sections and is similar in organization to the Yale University copy described in the Macphail *Catalogue*:[36] 1) engraved frontispiece, headed *Signaculum Mundi Pythagoricum* (see figure three); 2) engraved title page, dated 1673; 3) general title page with extended descriptions of the four major treatises, dated 1673 (see entry no. 105); 4) the Boyle dedication, dated 1 April 1671; 5) the Ashmole dedication, dated 16 January 1668; 6) the "Author's Epistle to the Reader," dated 10 January 1670; 7) the individual title page of the *Philosophical Epitaph*; 8) the double page, Latin and English, of illustrations headed: "*An* EPITAPH *made by* W.C." (figure two); 9) the four chapters of the *Epitaph* immediately preceded by a second appearance of the frontispiece. These sections are, in turn, followed by the texts of Helvetius's *Brief of the Golden Calf*, Glauber's *The Golden Ass well managed*, the anonymous *Jehior*, and the *Catalogue of Chymicall Books*.

The dedications to Boyle and Ashmole and the "Author's Epistle to the Reader" are all signed "W. C. or twice five Hundred," and followed by the anagram *L'Aurum Amice eligis Rus*; together these three brief prefatory sections provide most of the small stock of information that is available on the author. I begin with the one bearing the earliest date.

---

36 Macphail, *Alchemy and the Occult*, II, 445–50.

Engraved title page, probably by Robert Vaughan, of *The Philosophical Epitaph of W. C. Esquire, for a Memento Mori on his Tomb-stone* (London, 1673). Bodleian, Ashmole 1348. Reproduced with permission of the Bodleian Library.

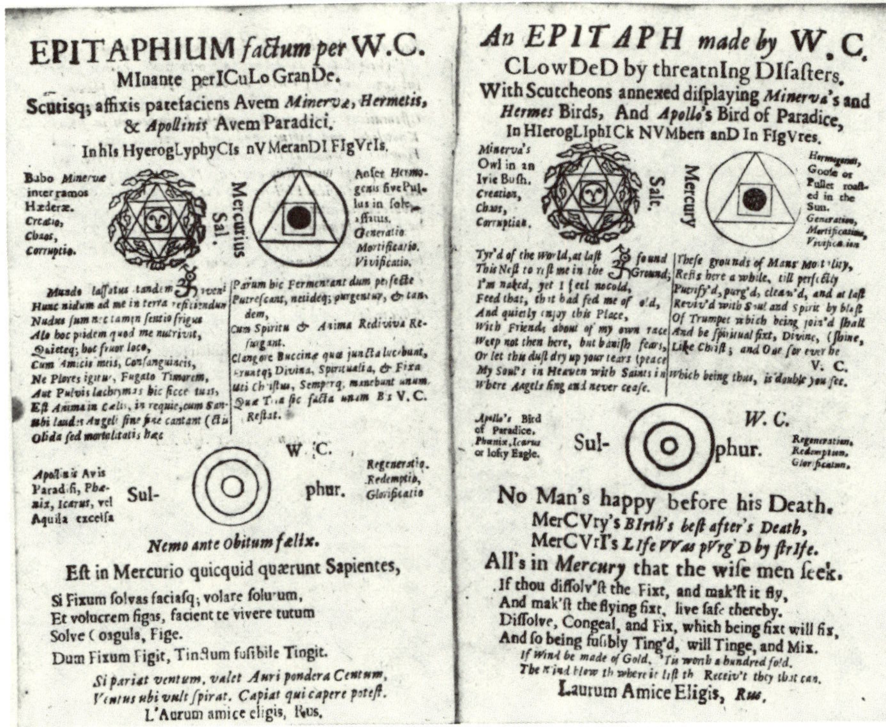

EPITAPHIUM factum per W.C.

MInante perICuLo GranDe.

Scutisq; affixis patefaciens Avem *Minervæ*, *Hermetis*, & *Apollinis* Avem Paradici,

InhIs HyerogLyphyCIs nVMeranDI FIgVrIs.

Bubo Minervæ inter ramos Hæderæ. Creatio, Chaos, Corruptio.

Mercurius

Sal.

Anser Hermogenis sive Pullus in sole assatus. Generatio. Mortificatio. Vivificatio.

Mundo lassatus tandem [symbol] veni
Hunc nidum ad me in terra refitiendum
Nudus sum nec tamen sentio frigus
Alo hoc pridem quod me nutrivit,
Quieteq; hoc fruor loco,
Cum Amicis meis, Consanguineis,
Ne Plores igitur, Fugato Timorem,
Aut Pulvis lachrymas hic sicce tuas,
Est Anima in Cælis, in requie, cum Sanctis
ubi laudes Angeli sine fine cantant (Elia
Olida sed mortalitatis hæc

Parum hic Fermentant dum perfecte
Putrescant, nitideq; purgentur, & tandem,
Cum Spiritu & Anima Redivivia Resurgant.
Clangore Buccinæ quæ juncta lucebunt,
crantq; Divina, Spiritualia, & Fixa
uti Christus, Semperq; manebunt unum.
Quæ Tria sic facta unum Bis V. C.
Restat.

W. C.

Apollinis Avis Paradisi, Phænix, Icarus, vel Aquila excelsa

Sul- phur.

Regeneratio. Redemptio, Glorificatio

Nemo ante obitum fælix.

Est in Mercurio quicquid quærunt Sapientes,

Si Fixum solvas faciasq; volare solutum,
Et volucrem figas, facient te vivere tutum
Solve Coagula, Fige.
Dum Fixum Figit, Tinctum fusibile Tingit.

Si pariat ventum, valet Auri pondera Centum,
Ventus ubi vult spirat. Capiat qui capere potest.
L'Aurum amice eligis, Rus.

An EPITAPH made by W. C.

CLowDeD by threatnIng DIsasters.

With Scutcheons annexed displaying *Minerva's* and *Hermes* Birds, And *Apollo's* Bird of Paradice,

In HIeroGLIphICk NVMbers anD In FIgVres.

Minerva's Owl in an Ivie Bush. *Creation, Chaos, Corruption.*

Salt.

Mercury

Hermogenes Goose or Pullet rosted in the Sun. *Generation, Mortification, Vivification*

Tyr'd of the World, at last [symbol] found
This Nest to rest me in the [symbol] Ground;
I'm naked, yet I feel no cold,
Feed that, that had fed me of old,
And quietly enjoy this Place,
With Friends about of my own race
Weep not then here, but banish fears,
Or let this dust dry up your tears (peace
My Soul's in Heaven with Saints in
Where Angels sing and never cease.

These grounds of Mans Mortality,
Rests here a while, till perfectly
Purify'd, purg'd, cleans'd, and at last
Reviv'd with Soul and Spirit by blast
Of Trumpet which being join'd shall
And be spiritual fixt, Divine, (shine,
Like Christ; and One for ever be
V. C.
which being thus, is double you see.

Apollo's Bird of Paradice. Phœnix, Icarus or lofty Eagle.

W. C.

Sul- phur.

Regeneration, Redemption, Glorification,

No Man's happy before his Death.

MerCVry's *BIrth's best after's Death,*

MerCVrI's *LIfe VVas pVrg'D by strIfe.*

All's in *Mercury* that the wise men seek.

If thou dissolv'st the Fixt, and mak'st it fly,
And mak'st the flying fixt, live safe thereby.
Dissolve, Congeal, and Fix, which being fixt will fix,
And so being fusibly Ting'd, will Tinge, and Mix.

If Wind be made of Gold, 'Tis worth a hundred fold.
The Wind blow th where it list th Receiv't they that can.

Laurum Amice Eligis, *Rus.*

"*An* EPITAPH *made by* W. C. CLowDeD by threatnIng DIsasters," from *The Philosophical Epitaph* (London, 1673). Bodleian, Ashmole 1348, before Sig. B. Reproduced with permission of the Bodleian Library.

Signaculum Mundi Pythagoricum, engraved frontispiece from *The Philosophical Epitaph* (London, 1673). Bodleian, Ashmole 1348. Reproduced with permission of the Bodleian Library.

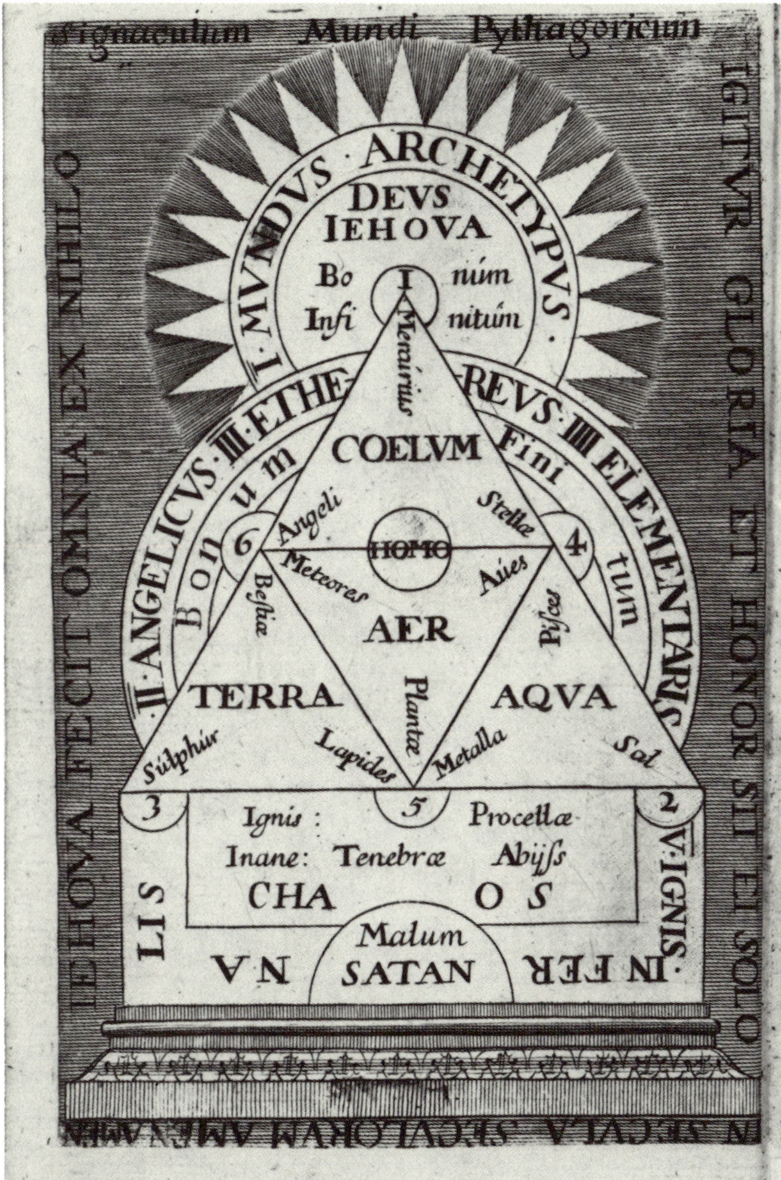

Title page engraved by Robert Vaughan for *Ripley Reviv'd*, by Eirenaeus Philalethes, printed for William Cooper, (London, 1678). Bodleian, 8°. S. 40. Med. Reproduced with permission of the Bodleian Library.

As reported by Plomer, Macphail, and others, the Ashmole dedication alludes to an earlier period of trouble when W. C. had been helped by Ashmole, having "received some civilities from you, . . . in the small intermission of my Long troubles, 1662." Cooper, or so I will refer to the author, then recalls an even earlier time of great illness during which the hieroglyphical "scutcheons" were designed:

> they were made in a Living Grave, 1652. From whence I never thought to come forth no more, then probably *Jonas* might in the Whales Belly, *Daniel* in the Lyons Den, or the three Children in the Fiery Furnace, Being grievously oppressed and clowded in my long Troubles, and since as little regarded.[37]

While Cooper provides no additional information on the nature of his "troubles" or illness, he does suggest why he may have dedicated his work to the great antiquarian. Early in the dedication, he refers to the fact that his work, initially intended only as a translation of the Helvetius tract, has been expanded to include the "*Epitaph*, with several Scutcheons, Mottos, and Explanations, with *Pythagoras* his *General Figure*, Blazoning Philosophical Herauldry, and also the *Alchahest, Samech'* with other *Elixirs, &c.*"[38] Thus the dedication was not only appropriate because of Ashmole's past kindness but because of his position as one of Charles II's heralds at arms. Ashmole's principle work, the *Institutions, Laws, and Ceremonies of the Order of the Garter* (1672), appeared after the dedication was penned but before the *Epitaph* was published. But there is another important reason for Cooper's solicitousness: he is well aware that Ashmole's interests in alchemy coincide with his own, and he is familiar with Ashmole's *Theatrum Chemicum Britannicum* (1652; entry no. 11): "those Worthy Collections of English Philosophick Chymical Authors, formerly published by you."[39]

The most interesting comments in the Ashmole dedication, however, are those which alert us as to what the hieroglyphical scutcheons

---

37 *Philosophical Epitaph* (London, 1673), "To His Worthy, and much Honoured Friend, Elias Ashmole Esq;" sig. A5–A5^{v}.

38 Ashmole Dedication, sig. A4–A4^{v}.

39 Ashmole Dedication, sig. A4^{v}. In the "Advertisement" preceding *Ripley Reviv'd* (1678), Cooper again commended Ashmole for producing "his most excellent and laborious Collection of our Ancient English Hermetick Authors." Moreover, he calls upon the "learned Philosophers" of the time to encourage Ashmole to publish the proposed second volume of the *Theatrum*, "which he had almost finished near twenty years since, (as I had it from his own mouth) and hath lain asleep ever since, and likely so to lie; for to the perfecting thereof he is now unwilling to be brought." Cooper's justifiable fear was that Ashmole would die before publication of the second volume of these "rare pieces of Antiquity."

and, more generally, the *Epitaph* mean and how they are to be interpreted. With an almost playful glance at Ashmole's interest in heraldry, Cooper states:

> I know you have another Coat of Arms for my Paternal Family, in your Heraulds Office, which would suit with the said Epitaph, if it were only as it is also intended for a plain Sculpture to be upon my Grave Stone: Yet such Scutcheons had not been so proper for this place, these being chiefly here intended for the Philosophers Stone, agreeing with my said Epitaph, in the Elements, Principles, and whole perfection thereof, excellently manifested by our late *English Phoenix*, or *Elias Artisto Anonymon* [i.e. Eirenaeus Philalethes], in his Book of *The open entrance to the shut Pallace of the King.* Now some perchance may think it incongruous for any man to publish his own Epitaph, or annex any such Novel Scutcheons. Yet since they and their Explanations are Philosophick [i.e., alchemical] . . . I was the more confident of allowance and approbation.[40]

The subject of hieroglyphical interpretation is touched on in the second and later of these prefatory statements, "The Authors Epistle" of 10 January 1670. Cooper repeats his previous assertion that he had not originally intended to publish the "Figur'd Scutcheons" other than on his grave stone; however, and the strain of conventional alchemical piety is present in each of these addresses, his life has been spared by Almighty God and he has been "providentially" provided with the "adjoyned Treatises for my Task before I dyed." Therefore he has determined to interpret these designs for the praise of God and the betterment of his fellow men:

> I thought good to offer my mean Mite to the World, so that thou mayest not only see and read an AEnygma in these my Scutcheons and Epitaph, but have me thy AEdypus to unfold them. Where also I have set forth the Philosophers Stone, and shewed the Causes and manner of Multiplication of Life and Seed . . . with other Elixirs, and Philosophick Medicines, &c. in 5 small succinct Chapters.[41]

The dedication to "Robert Boyl[e] Esq; Eminently Noble & Accomplisht" is the last and latest of the three prefatory addresses, dated 1 April 1671, nearly three years after the dedication to Ashmole and two years before actual printing of the *Epitaph*. The reason for this

---

40 Ashmole Dedication, sig. A4^{v}–A5.

41 "The Authors Epistle to the Courteous and Well minded Reader," sig. [A].

delay is explained at the beginning: W. C.'s translation of Helvetius' *Golden Calf*, to have been the only part according to the original plan, was licensed for printing as early as 1668. Before it could be published, however, another very similar translation appeared,[42] thus obviating the need for his own. W. C.'s report of the matter is confirmed by the *Stationers' Register* entry for 29 September 1668, which registers to Master William Cooper "his copie . . . [of a] booke intituled *Vitulus Aureus, or the Goulden Calfe*."[43] In view of W.C.'s next statement in the Boyle dedication, there can be no doubt that the translation of Helvetius included in the *Epitaph* is based on the earlier one, the printing of which had been aborted:

> Nevertheless it [the postponement] hath brought advantage to the Reader, for I have since exceedingly abbreviated my former Translation with the Epistles, &c. not diminishing sense or matter, and have adjoyned my own Philosophical Figured Epitaph.[44]

Given the fact that the first, aborted translation was referred to in the *Stationers' Register* as Master William Cooper, "his copie," there can be little doubt that W. C. is Cooper, particularly when we add to the case the other evidence that has been adduced.

In chapter one of the *Epitaph* proper, entitled "A plain and full explanation of the aforesaid Epitaph, Scutcheons and Motto's of W. C. As well for the Philosophers Stone as his own Tomb-stone," Cooper, as he had promised in the "Authors Epistle," sets out to be the AEdypus who will "unfold" the "philosophick" meaning of these enigmatic designs. Thus, the title serves to place the chapter and the *Epitaph* generally in the tradition of analogy, esotericism, and doubleness of vision that is common to alchemical writing. In the seventeenth century, in particular, these qualities often combine with fervent piety and moral seriousness to produce a mode of expression that professes plainness and accessibility even as it depends upon a discourse marked by vagueness, ambiguity, obscurity, abstractness, and private symbolism. Cooper's alchemical writing shares in all of these weaknesses and is, as a representative of this genre, undistinguished; nonetheless, the *Epitaph* is not without points of interest.

If chapter one purports to be a gloss on the scutcheons and mottos represented in figure two, it is more narrowly a gloss on one particular

---

42 Boyle Dedication, sig. A2–A2^v. This translation was *The Golden Calf, which the World adores . . . written in Latin by John Frederick Helvetius . . . and faithfully Englished* (London, 1670). (See entry no. 188.)

43 *Stationers' Register*, II, 391.

44 Boyle Dedication, sig. A2^v.

symbol: the human-like composite of several planetary signs that appears, on both the Latin and the English pages, embedded in the lines of verse just below Minerva's owl. I have argued elsewhere that this figure, which was variously employed by a small number of seventeenth-century writers, is ultimately derived from John Dee's *Monas Hieroglyphica* (1564) where, interpreted mystically and cabbalistically, it represents alchemy.[45] For W. C., its meaning is both greater and less. On the one hand, the figure represents the author and, in keeping with the tract's moral intent, the ubiquity of death; on the other, it represents the entire alchemical process and its chief ingredient, the Mercury of Philosophers, which like the Christian, must suffer death as preparation for regeneration:

> This Epitaph is literally the work of Philosophers, and yet may revive the old useful Adigy and Motto upon this Authors Tomb-stone, to remember thy end. For as this flourishing [symbol][46] signifies this Author *W C.* being a Mercurialist, tired of all worldly Inquinaments So it illustrates all the Planets and their Mercury, and the universal Spirit and Mercury of the World, . . . and no less, the true Mercury of Philosophers for this work: free from all filthy corruptions well fitted and put naked without Garb, or any strange thing into its Glass, and private philosophical Nest or Vessel (as into a Grave and Coffin) with constant vapourous heat for Putrifaction, and its true preparation, rectification, and perfection.[47]

From this opening, the first chapter proceeds to summarize many of the ideas and principles that are commonplace in alchemical tradition. We see, for example, that the preparation of the philosopher's stone can be described in terms of a series of complex analogies which are predicated on numerological equivalencies: greatly simplified, a duality of principles (e.g. male and female, or sulphur and mercury) may be divided into three parts, such as body, soul, and spirit, or salt, sulphur, and mercury. (The general orientation of the *Epitaph* is toward the Paracelsian *tria prima.*) Because the alchemical process rearranges the qualities that constitute the elements, it is necessary to move from the triplicity of the parts to the quadraplicity of the elements and the four degrees of the alchemists' fires. Finally, these numerologically-conceived principles and processes "at the last will make a fifth Essence; and so by a perfect Ternary *Quadrate*, and Quintessential Process, from one, two,

45 See my "Robert Burton's Sigil and John Dee," *Notes & Queries* 29 (October, 1982), 415–16.

46 Here Cooper inserts the aforementioned composite symbol.

47 *Philosophical Epitaph*, p. 1.

three, four and five." W. C.'s numerological inclinations even extend to his initials: earlier it was noted that the prefatory addresses are signed "W. C. Or twice five Hundred." In chapter one he explains both the literal meaning of the analogy, i.e., V V C=5 plus 5 x 100=1000, and its "philosophick" signification, with respect to the potency attained by the great "Medicine" or elixir which is the end of the alchemist's quest:

> Thus then at last this Medicine may obtain the name and number, intimated by *W. C.* which as it is this Authors name, who is but one in Person, and in Figures, twice five hundred; so is the Medicine but one in substance, and in virtue twice five hundred, or a thousand.[48]

Our chief difficulty in the *Epitaph* arises from the attempt to interpret the symbolic scutcheons and mottos by means of the gloss of the text in chapter one or, conversely, to use the illustrations as aids in clarifying and visualizing the printed word. In Cooper, the visual and verbal elements lack the congruity and mutual reinforcement that we typically find in an emblem of Alciati, Quarles, or Wither, and this, combined with the obscurity of the subject matter, results in ambiguity. It is clear, nevertheless, that W. C. is intent on emphasizing the progressive stages of the alchemical process in both the symbolic scutcheons and the text: each of the three designs comprehends, for example, a triad of alchemical processes: Minerva's Owl (creation, chaos, corruption); Hermogenes' Goose or Pullet (generation, mortification, vivification); and Apollo's Bird of Paradise (regeneration, redemption, glorification). These designs are, in turn, coordinated with short sections of verse within chapter one which purport to elucidate the diagram, as in the following example:

> The Chaos in th'Excentrick Centre still,
> Hath death's Heads Ternary, Crows or owly Bill.
> Whose square Face, under Times confused Glass,
> Of Fire and Water, six days Angles pass,
> Within the Spiny Bush, Expansion till,
> A Sabatean Rest makes all stand still.
> After each Colour fram'd to th' owners praise,
> Then all things multiply to the end of days.
> The two in number, are but one in kind,
> And four in Nature, three in one do bind.
> And then the Quintessence wheels thrice in'ts Sphear,
> To conquer all the Mortals every where;[49]

48 *Epitaph*, p. 3.

49 *Epitaph*, p. 5.

Here, the references to the "death's Head," "owly Bill," "six days Angles," "Spiny Bush" and to the numbers (six, two, one, four, three, and five) and the geometrical shapes (the circle, square, and hexagon) point to details of the design, and thereby attempt to clarify its alchemical meaning. The "chaos" or alchemists' *prima materia*, sealed within the alembic for six days, undergoes, through the agency of fire and water, a process of "death" or blackening, becoming the color of the crow. This state of death or inertness is illusory, however, and following a "Sabatean Rest," a procession of colors appears, marking the beginning of the regenerative phase. It is the culmination of the latter, the "Quintessence" or philosopher's stone, that has the potency "all things [to] multiply." While this alchemical meaning is being unfolded by illustration and text, the verses beneath the diagrams impart the personal, redemptive message that is closely, if arcanely, linked to the alchemical. On the basis of such analogies W. C. is able to claim that "as this Epitaph doth thus set forth the true Elixir of Philosophers, and mans Mortality; so likewise these Scutcheons or Hyeroglifical Figures you see do the same in the honourable Pedegree of the Philosophick true Medicine, . . . as well for the Life and Health of mans Body as Metalls."[50]

In other respects, as well, the *Epitaph* is quite conventional: it recognizes the diversity of opinion that exists concerning the preparation of the philosopher's stone and is complete with lists of authorities (e.g. St. Dunstan, Arnold of Villanova, Raymund Lull, Sir George Ripley, and Glauber). Like many English treatises of this period, it is Paracelsian in its assertion of the three hypostatical principles of salt, sulphur, and mercury; and its idea of alchemical transmutation, perfection, and healing—in humans no less than in metals—is grounded in the notion that the philosopher's stone is Christ-like in its regenerative potency. This we see in the imagery which W. C. uses, at the conclusion of book one, in describing the millenarian-like coming of the great elixir:

> And then the Glorified King (*Triply* Crown'd) shall vanquish his Enemies, and redeem his Brethren and Kindred, in all or any Nations from their vile Corruptions: If they can but touch the hem of his Garment; or entertain him at his approach, as they ought; for'tis alike to him, to raise their Essences, as to separate their Maladies.[51]

---

50 *Epitaph*, p. 4.

51 *Epitaph*, p. 10. A useful study of the Christ-philosopher's stone correspondence appears in C. G. Jung, *Psychology and Alchemy*, trans. R. F. C. Hull, 2nd ed. (Princeton: Princeton Univ. Press, 1953), pp. 345–431. See also Thomas S. Willard, "Alchemy and the Bible," in *Centre and Labyrinth: Essays in Honour of Northrup Frye*, ed. Eleanor Cook, et al. (Toronto: Univ. of Toronto Press, 1982,), pp. 115–27.

While the *Philosophical Epitaph* is distinctly Christian in its tone and imagery, it is Christianity at times strikingly and significantly infused with Neoplatonic ideas and images. Particularly in chapter two, "Of the Causes and Manner of Multiplication of Life and Seed," W. C. draws heavily on Plotinian and Ficinian doctrines of creation and form, universal heirarchy, the penetration (or as he would have it, "conversation") of the intellectual and the elemental worlds, and above all, of the *anima mundi*. This emanation of Nature, which operates through the forms and seeds implanted by God at creation, controls and orders the natural world. The interplay of the Christian and Neoplatonic is important enough to quote at some length:

> In the beginning God gave his blessing to increase and multiply; and commanded that each Thing from its like, should draw its Form; and so created in Nature a certain Chain, or subordinate propinquity of Complexions, between Visibles and Invisibles; by which the Superiour Spiritual Essences descend, and converse here below with the matter. Yet nature hath, nor had but one onely Agent (hidden in the universe) which is *Anima Mundi*, working by its universal Spirit, through innumerable distinct, Concreates according to their Specificq [sic], Forms and Seeds, which God the Father at first Creation by his word and Idea (or Son and Holy Spirit) did glance at once into the first matter, and so set Laws and Bounds in Nature: . . . so that a Sparrow falls not without his Providence and Power; . . . . There is no part of Nature which doth not contain within it self the seed of its own kind.[52]

W. C.'s comment on fate and God's providentiality ("a Sparrow falls not," etc.) is also predicated on ideas that are more Plotinian than Judeo-Christian. The production of metals beneath the earth, he relates, is the result of a combining of Sulphur and Mercury in which accident has no part. Rather, the conjunction and its resultant metal, comes by "Gods Decree; Providence and Power, from its Specificq [sic]; Seed, and its hidden inward Agent, Form and Life, [and] from *Anima*

52 *Epitaph*, pp. 11–12. Neoplatonic analogues may be seen, for example, in the *Enneads* I, Tractatus 6, where Plotinus speaks of the relationship between the beauty of the One and that of the elemental world. "Ideal form," emanating from the Divine and descending through the lower hypostases, is received by the individual material object: "Let us, then, go back to the sources, and indicate at once the Principle that bestows beauty on material things. But, is there any such likeness between the loveliness of this world and the splendours in the Supreme? Such a likeness in the particulars would make the two orders alike; but what is there in common between beauty here and beauty There? We hold that all the loveliness of this world comes by communion in Ideal-Form." Beauty is thus what the Soul recognizes as a congruence between Divine Ideal and the particular. *The Enneads*, trans. Stephen MacKenna, 2nd ed., rev. (1917–30; rpt. London: Faber and Faber, 1956), pp. 57–8.

Mundi."[53] Throughout the *Epitaph*, W. C. affirms the efficacy of this seed or light, implanted by God, to shape and control the development of the material object from within.

Chapters three and four are both concerned with the preparation of elixirs and universal medicines, such as the *alchahest*, and are obviously indebted to Paracelsus, "the glory of Chymists" and "brave *Helmont* his great Interpreter."[54] In fact, he acknowledges at the outset of chapter five that his work is intended as a key to "open" these two authors, to expose them in their "concealments." This short final chapter concludes, however, on the conventional notes of piety and humility. If he desires the secret of the philopher's stone, the reader must pray to God and prepare himself in the ways of "purity and holiness." The result of such preparation, if the petitioner receives God's blessing, will be illumination and regeneration through implantation of "an Essential and Living Seed of Faith." Thus perfected, he will "mount through and above the quaternary defiling world into the Triune power, and at last come to the quintessential, or Super celestial Central circle of Peace, and Heavenly Beatitude."[55] Thus eternal rewards are conceived more along the Plotinian lines of the soul's ascent through the lower hypostases to the One than in accordance with the plainer truths of Christian orthodoxy.

Though, as noted earlier, much alchemical lore included in the *Epitaph* is conventional, Cooper's efforts to coordinate "philosophick" ideas with a personal and poetic *memento mori*, and to draw together visual illustration and text into a mutually illuminating congruity, as well as his unequivocal references to Neoplatonism in chapter two, do give the work a measure of originality. While other details of the treatise could be profitably investigated, brief attention must be given to a few of the personal utterances that Cooper includes in the later sections of the *Epitaph* volume. We have seen that Cooper's original intention was to publish a translation of Johann Friedrich Helvetius' *Golden Calf*, that this intent was frustrated by the appearance of an edition printed by John Starkey in 1670 (Wing H1410), and that Cooper then settled upon the idea of publishing a shorter version of Helvetius (*A Brief of the Golden Calfe*) together with his own *Philosophical Epitaph* and the other treatises which make up the 1673 volume.

---

53 *Epitaph*, p. 13. Again, Cooper is clearly drawing upon the seed imagery so popular in Neoplatonic philosophy. To cite only one example, Marsilio Ficino's *Commentary on Plato's Symposium*, Speech VI, chapter 12, states: "The seeds of all of the body's own things are implanted in it from the beginning. Hence, at appropriate intervals of time the teeth come through, hair appears, a beard develops, and the seeds of procreation begin to flow," Sears Jayne, tran., *Commentary on Plato's Symposium on Love*, 2nd, rev. ed. (Dallas: Spring Publications, 1985), p. 132.

54 *Epitaph*, p. 17.

55 *Epitaph*, p. 20.

The Helvetius translation is prefaced by "The Epistle of W. C. to the Reader," undated but signed in the customary cryptic way. Here, more than in any other of the personal statements we have considered, Cooper appears to regard the art of alchemy very seriously, in fact to believe in "the reality of the Philosophers Stone & Universal Medicine, and consequently to esteem the Noble Art of Chymistry by which it is wrought."[56] He deplores the strictness of the laws that discourage "honest Practisers of this Art." At the same time he applauds publication of significant works like the *Golden Calf*, Ashmole's *Theatrum Chemicum Britannicum*, and English translations of Glauber, Basil Valentine, Sendivogius, Paracelsus, and, in particular, the Philalethean *Secrets Reveal'd* printed for him by William Godbid in 1669. With perhaps a hint of self-congratulation and economic self-interest, he notes that "There is also published a Manuscript of a most rare *Anonymon* (probably yet living) who like a miracle of nature, attained the *Elixir* at 23 years of Age, 1645. And as a true *Elias* (or fore-runner) hath taught the same, in his Book Entituled, *Secrets Revealed, or an open entrance to the shut Pallace of the King.*"[57] These publications serve to bring "honour [to] this Art, Improvement of our English Nation, and to establish a belief of the said Stone." He cites the case of Helvetius himself, who had at one time been sceptical of the art, but had been won over through witnessing demonstrations of its validity. Cooper's own acceptance of alchemy is reflected in his statement to the reader that "if these Experiments shall gain the like credit with you (as I doubt not but they may) you will not any ways detract or scandalize this almost Divine Art."[58]

Cooper's address concludes with further examples of the three tendencies which marked the end of the *Philosophical Epitaph*, the expression of piety, religious allegorizing, and millenarian anticipation, all highly characteristic of alchemical treatises written in England in the mid-seventeenth century. The truth of alchemy, he states, should teach us to "magnifie the great Creator, who hath not only given us this pretious [Philosopher's] Stone for our health and wealth, but withal a most glorious white Stone, clothed in Scarlet, *viz.* his Son Christ Jesus for the Example, Redemption, and Eternal Salvation of all men of that Spirit."[59] Further, he expresses hope that the public revelation of alchemical secrets will foreshadow an era of universal spiritual and social reform to coincide with Christ's reign on earth. Cooper's appeal

56 "The Epistle of W. C. to the Reader," prefacing *A Briefe of the Golden Calf.*

57 "Epistle . . . to the Reader."

58 "Epistle . . . to the Reader."

59 "Epistle . . . to the Reader."

characterizes the alchemical millenarianism of his age:

> But methinks these bright Stars thus eminently appearing, with other manifest Tokens would perswade us that the time is come, or not far off, when the true *Elias* is or will be revealing this and all other Arts and Mysteries more plainly and publickly then before, though not perchance in or by any single person, but in some publick Administration of Spirit (like a second *John Baptist* in a Fiery Chariot) to prepare the way for a higher design, by which men may forsake their vain lusts and pleasures, to follow this and other laudable Arts. . . . till they come to be transformed into the perfect Image of Christ, in, by, and with whom he will Reign spiritually.[60]

Again, in "The Epistle of W. C. to the Christian and Courteous Reader" that precedes his translation of J. R. Glauber's *Golden Ass Well managed, and Midas restor'd to Reason*, Cooper emphasizes the religious and moral benefits that may accrue, even in these "Calamitous times," from reading the treatise and from successful practice of the art. Though he freely admits that alchemy may be a means to wealth and power, the Art should properly "breed in thee a holy hunger of God, rather then Gold, and improve this Talent to Gods Honour that sent it, and to thy honest Neighbours good."[61] As a result of this spiritual transformation, he concludes in a millenarian vein, "the true Golden Age, so long expected and desired" may arrive, a time when all will live together harmoniously. That Cooper is coming to believe seriously in alchemy is perhaps further reflected in the fact that a Postscript, immediately following this Epistle, provides directions for making a concoction that will produce "good profit."[62]

The last of the prefatory epistles, which precedes the final treatise in Cooper's volume, is addressed to "the honest, sober Reader" and dated 3 July 1672. While it is of some interest as an introduction to *Jehior or The Day dawning; or Morning Light of Wisdom* and is again marked by piety and millenarianism, it also provides insights into Cooper's rationale for producing the entire volume. Noting that *Jehior*

---

60 "Epistle . . . to the Reader." On the Jewish and Christian mystical tradition of the return of Elias as a foreshadowing of Christ's Second Coming, and its influence upon Renaissance Utopian, Rosicrucian, alchemical and scientific thought, see Herbert Breger, "Elias Artista—A Precursor of the Messiah in Natural Science," in *Nineteen Eighty-Four: Science Between Utopia and Dystopia*, ed. Everett Mendelsohn and Helga Nowotny (Dordrecht, Holland: D. Reidel, 1984), 49–72. See also my "Alchemy and Eschatology in Seventeenth-Century Poetry," *Ambix*, 31 (November, 1984), 102–24, especially note 53.

61 "The Epistle of W. C. to the Christian and Courteous Reader."

62 "The Epistle," Postscript, sigs. G–G^{v}.

"was published some years since" but was now out of print, Cooper states that since the tract had in the interim been "improved by the Author" (usually taken to be Paul Felgenhauer) and was "sutable to Pythagoras his Metaphysical and Physical Figure, with my smaller *Philosophical Epitaph and Figures*, I thought good to make them with the rest into one small Volume, where much light of Divinity and Philosophy will appear, concentrated and multiplied to any ingenious Spirits."[63]

I have noted earlier that in chapter two of the *Philosophical Epitaph*, Cooper's Christianity is strongly colored by Neoplatonic ideas and imagery. It is undoubtedly this aspect of his religious and philosophical temperament that responded to the *Jehior*, caused him to select it for inclusion in his volume, and enabled him to see the correspondences between this treatise and the Pythagorean "Metaphysical and Physical Figure" which he had already decided to include as an illustration. The *Signaculum Mundi Pythagoricum* (figure three) which appears both as a frontispiece and an accompaniment to the *Jehior*, does not, however, originate with the *Philosophical Epitaph*; it is ultimately derived from an illustration that appeared in Helisaeus Roslinus' *De opere dei creationis seu de mundo hypothesis* (1597), though, as S. K. Heninger, Jr. has noted, Cooper is likely to have used the more recent version published in the *Musaeum hermeticum* (Frankfurt, 1625) as the immediate source.[64] While Heninger's explanation is essential for a full understanding of the diagram's complexity, some brief commentary will at least suggest connections between the diagram and the *Jehior* and with Neoplatonism generally.

In the diagram, the (I.) archetypal idea of the universe, existing in and emanating downward from the mind of Jehovah, the One who is *Bonum Infinitum*, is manifest in the creation of the three "lower" heirarchically-arranged spheres: (II.) the Angelic, (III.) the Ethereal, and (IV.) the Elemental. Reflecting the pattern in the mind of their creator, who is the center and source of radiating light, these spheres are good, but finite. In the middle of the elemental world (and in the midst of the Paracelsian *tria prima* of salt, sulphur, and mercury) is man, partaking of both the celestial and terrestrial spheres. Below is the (V.) infernal realm of Satan, chaos, and fire.

Reflecting Plotinian and Proclian fondness for analogy and triadic

---

63 *Jehior or the Day dawning; or Morning Light of Wisdom,* "The Epistle to the honest, sober Reader."

64 My comments on this figure are heavily indebted to the discussion of S. K. Heninger, Jr., *Touches of Sweet Harmony: Pythagorean Cosmology and Renaissance Poetics* (San Marino: Huntington Library, 1974), pp. 158–59, and *The Cosmographical Glass: Renaissance Diagrams of the Universe* (San Marino: Huntington Library, 1977), pp. 126–27, 200.

structures as well as familiarity with the opening verses of the gospel according to St. John, the author of *Jehior* speaks of the Word, "the true Principle in God himself," as that from which all proceeds and which is particularly manifest in the three principles of God, Nature, and the Elements.[65] It is from these principles, and here Cooper's "Pythagorean" diagram is especially "sutable," that a three-fold world was created:

> namely a divine uncreated from God flowing . . . from Eternity, then an Angelical world, which proceedeth or lighteth forth, or shineth forth out of the Light in which God dwelleth; and lastly an Elementary world, whose Original came out of the water.[66]

As the Angelic and Ethereal spheres are in an intermediate position between God and the elemental sphere according to the diagram, so Nature occupies this position for the author of the *Jehior*: it is "the second Principle and beginning of all things, and stands betwixt God and the Elements, through which God worketh into the Elements." Or again, he posits Nature as an emanation, a "flowing out" from the divine world, just as "the Elementary world is an overflowing of the Natur[al] and Angelical world." Finally, and in terms that are unmistakably Neoplatonic, "man cometh forth out of the three worlds, and is the concentrated or conjoyned *Centrum* of all the worlds."[67]

The *Jehior* abounds in other Neoplatonic images and ideas too numerous to discuss in detail. In chapter three, the conception of God, the first principle, is strongly numerological and Pythagorean: "the Mystery of God consists in One, Three and Seven; and according to this Mystery all other things are Created and consist in One, out of Three, through Seven."[68] The later discussion of the Paracelsian principles of salt, sulphur, and mercury is self-consciously Proclian in its triadic multiplications and oddly attenuated analogies to the body, soul, and spirit.[69] *Jehior's* creation account incorporates the images of seeds and light, recalling Plotinian and Ficinian doctrines: "But the spirit of the Lord moved upon the water, thereby it became *seedy*, and the first thing that was created in it was *Light*."[70] Finally, the milieu that is evoked by the author of the *Jehior* is both magical and millenarian: Daniel and the prophets, he asserts, had studied magic "and went beyond the wise

65 *Jehior*, p. 5.

66 *Jehior*, p. 6.

67 *Jehior*, p. 16.

68 *Jehior*, pp. 10–11.

69 *Jehior*, see, for example, pp. 16–18.

70 *Jehior*, pp. 48–9.

men & magicians of *Egypt* far with their skill."[71] But the present time is also mysteriously ominous:

> Now is a very dangerous time; because all spirits are stirring; because their end is so nigh, that it is hardly believed. The air is full of spirits, and the Earth also is full of them, and every man hath his proper Angel and his bad Angel also.[72]

The Catalogue of Chymicall Books

Following the *Jehior* and included as the final section in the *Philosophical Epitaph* volume of 1673, Cooper added the *Catalogue of Chymical Books which have been written Originally, or translated into English.* The shortest of three versions and printed without prefatory addresses or explanations, the 1673 edition includes twenty-five pages (unpaginated but with signature marks as follows: P, P2, Q, Q2, R, R2) and approximately 186 main entries, excluding both cross references and those for different editions of a given title. Authors or, lacking these, titles are generally and imperfectly alphabetized—tending to be grouped by letter rather than carefully and correctly alphabetized—and blank spaces have been left after most of the alphabetical groupings to provide room for additional entries. Within entries, Cooper uses initials and abbreviations freely and generally includes the place and date of publication and the volume size. Titles are usually presented in a shortened and descriptive form, and in the case of collections of treatises, Cooper often provides useful listings of individual authors and titles contained in a volume. Thus, a typical "complete" entry in the 1673 edition reads as follows:

> Don Alexis *of* Piemont, His *Collection of Secrets*, with the manner of making Distillations, &c. *Lond.* 1580. 4°[73]

The 1673 version provides the basis and format for the two later editions, those of 1675 and 1688. Though published with its own title page, i.e., *A Catalogue of Chymicall Books. In Three Parts*, and in some instances found independent of the *Philosophical Epitaph*, Cooper appears to have intended that the expanded *Catalogue* should accompany

71 *Jehior*, p. 55.

72 *Jehior*, pp. 36–7.

73 *A Catalogue of Chymical Books which have been written Originally, or Translated into English,* included in *The Philosophical Epitaph* (London, 1673), sig. P.

a reissue of the earlier volume. In the "Bibliopola Lectori" that prefaces the 1675 *Catalogue*, he explains to the "Gentle Reader" that the earlier catalog had been produced in great haste and "by way of an Essay," the result of his eagerness to "adjoyn it to the *Philosophical Epitaph*." He adds parenthetically that the *Epitaph* was "(then Printing for me, and unto which this [i.e., the expanded *Catalogue* of 1675] is likewise to be added)."[74] Again, in the brief explanation that precedes the 1688 version, Cooper reports this as a *Continuation or Appendix to the Second Part of the Catalogue* "printed with" *The Philosophical Epitaph*.[75] Thus while there appears to be no bibliographical necessity for the two works to have been published together, this was clearly Cooper's intention. The first part of the version of 1675 is, in any case, an exact reprint of 1673. However, Cooper tells us in the preface that in the interim he has "had longer time to view and peruse all those Books" (except for nine which he mentions by name),[76] and these together with many others have been included in the new "Second Part." Careful bibliographer that he was, Cooper also notes that besides the added books, he has "given a larger and more satisfactory account of the *Titles*, and several *Editions* of many of those Books mentioned in the *First Part*, unto which I often refer."[77] The characteristic direction that such expansions take is easily seen in the 1675 version of the Alessio entry previously cited:

> Don Alexis *of* Piedmont *his Collection of* Secrets, *in four Parts.* Lond. 1562, 1568. and 1595. *See more in the first Part of this Catalogue of Chymical Books, Printed with the Philosophical Epitaph.*[78]

Finally, Cooper mentions that the third part of the 1675 edition will be given over to an index, for a ten-year period, to the *Philosophical Transactions* of the *Royal Society* which have a bearing on "*Chymistry*, or the Study of *Nature* by *Art* in the *Animal*, *Vegetal* or *Mineral Kingdoms*. While the "Bibliopola Lectori" is unsigned, Cooper concludes it with the now familiar appeal to readers as potential purchasers:

> I likewise take the boldness to inform the *Reader*, that most of the Books contained in these *Catalogues* (with many others of this

74 *A Catalogue of Chymicall Books. In Three Parts* (London, 1675), "Bibliopola Lectori," sig. *2. At times, however, the 1675 *Catalogue* was issued independently of the *Philosophical Epitaph*; see Ferguson's note that this version appeared in a "thin pamphlet," (I, 135).

75 See the title page of the 1688 version, sig. D*2.

76 "Bibliopola Lectori," sig. *2.

77 "Bibliopola Lectori."

78 *The Second Part of the Catalogue of Chymical Books*, sig. A.

> Subject in *Latine*, a large account of which is given by *P. Borellus* in his *Bibliotheca Chymica* in 12°) are to be sold by *Will. Cooper* at the sign of the *Pelican* in *Little Britain, London.*[79]

While the 1675 version of the *Catalogue* includes much material that was not present in the first edition, Cooper does not suggest that the later version is complete and definitive; nor does he claim that his list has been compiled without the assistance of others. In the brief "Bookseller to the Reader" that immediately precedes the second part of the *Catalogue*, he states:

> nor do I pretend to publish this as an absolute Collection of *English Chymical Authors*; (distrusting I may have forgotten some of common note) but rather as an Essay to provoke others (better able) to perfect it. Several of these Books I have drawn out of the Catalogues of *And. Maunsel, William London*, and the *Mercurius Librarius*: others I have more largely transcribed from the Books themselves, with the Date when Printed, and in what Volume, as near as the shortness of my time would permit, having but a few days to Collect it.[80]

Despite this acknowledgement, it is difficult to judge with certainty the degree of Cooper's indebtedness to the earlier bibliographers, Andrew Maunsell and William London. Maunsell's *Catalogue of English Printed Books* (London, 1595) was compiled to put Queen Elizabeth's "most faithfull and loving subjects . . . in remembrance of the workes of so excellent Authors, namely, such books as have been either written or translated by godly & learned *Kings, Queenes, Martyrs*, and most noble *Confessors* of the truth of the word of God."[81] It is organized in two parts: works on divinity, subdivided into several heads, and those related to various scientific fields. These include the mathematical sciences, such as arithmetic, geometry, astronomy, astrology and music, and also the "Arte of Warre" and navigation. He also includes sections on medicine and surgery which demonstrate that here "our [English] tongue is not barrain."[82] Approximately twenty-three titles in Cooper's list also appear in Maunsell, but of these only two are quite likely to have been derived directly from the earlier catalog rather than from information in the books themselves. *A Letter sent by a Learned Physitian to his friend*

---

79 "Bibliopola Lectori."

80 "Bookseller to the Reader," before sig. A.

81 Andrew Maunsell, *The First Part of the Catalogue of English Printed Books* (London, 1595), "To the Queenes Most Sacred Maiestie," verso.

82 Andrew Maunsell, *The Seconde parte of the Catalogue*, "To the Right Honourable, Robert Earl of Essex and Ewe," verso.

(entry no. 234) and *A Profitable Discourse* (no. 312) are nearly identical to Maunsell's citations and are limited only to the information provided in the Elizabethan booklist. The fact that neither work is listed in the *Short-title Catalogue* or included, to the best of my knowledge, in the *British Library Catalogue* or that of any other major library, suggests that Cooper very probably derived his entry from Maunsell. While it is tempting to suppose that Cooper would have depended heavily on Maunsell for authors and titles of older scientific books, it is impossible to prove that this was the case.

Although William London's *Catalogue of the most vendible Books in England* (London, 1658) has many titles in its sections on "Physick and Chyrurgery" and "Books Mathematicall &c." that Cooper also included, it is unlikely that his compilation is heavily indebted to this source. Cooper's titles tend to be much fuller in their descriptiveness than London's and are usually supplied with publication dates, whereas London includes only authors, titles, and sizes. If Cooper made use of London he would also have to have gone to other sources to get the information that he normally includes. In contrast to the *Catalogues* of Maunsell and London, which are merely listings of factual information, Cooper's work is unique in that it reveals a high degree of familiarity with the authors and books it includes.

This intimacy with his material is reflected in many ways: in the fullness of detail present in his long descriptive titles, in his practice of cross-referencing individual sections within long works, such as the *Theatrum Chemicum Britannicum*, as well as in his reference to "Jam. Hasolle, *alias* Elias Ashmole." Even Cooper's errors may be seen as careless slips resulting from hastiness in working with familiar materials: the fact that he attributes *The Marrow of Chymical Physick* to George Starkey in the 1673 version rather than to William Thraster (or Thrasher) is a natural mistake because he is thinking of the *Marrow of Alchemy* which, according to Cooper, Starkey did publish. Again, the fact that he attributes Gabriel Plattes' *Discovery of Subterraneall Treasure* (entry no. 302) to Sir Hugh Platt, or that Henry More's well-known *Enthusiasmus Triumphatus* is referred to as *Atheismus Triumphatus* (see entry no. 264) can be seen as careless mistakes (usually corrected) committed by a busy man whose mind was making associations between authors and titles which could easily be confused. Fortunately, such errors are rare in the *Catalogue*, and as one works through the entries one is increasingly impressed with his knowledge of the subject and his accuracy in recording this information.

The third and final version of the *Catalogue*, dated 1688, is in some ways the most important, certainly the most puzzling bibliographically, and unquestionably the least well known. This edition bears the following title:

> *The Continuation or Appendix to the Second Part of the Catalogue of Chymical Books, Printed with The Philosophical Epitaph, being either Corrections, Rectifications, or Additions to the Editions of the old Books; and a Continuation of such New Books as have been Printed since the Year 1675. to this present Year 1688.*

Though not previously entered or noted in the *British Library Catalogue*, the 1688 version is bound into one of two British Library copies of the three-part *Catalogue* (shelf-mark 270. b. 16), having been inserted between the second and third parts. This *Continuation*, which occupies fifty pages marked D*2–D*26, appears to have escaped the notice of all scholars and bibliographers who have dealt with Cooper, and its rarity is evidenced by the fact that in addition to the British Library copy, only two others (located in the Library of Congress and the National Library of Medicine at Bethesda, Maryland) have thus far appeared.

The question of why so few copies of the *Continuation*, existing either separately or bound into the three-part *Catalogue* of 1675, are now extant cannot be answered with certainty. On the one hand, the major problems in the collation of this appendix, to which I will turn shortly, may suggest that it was an afterthought for Cooper or a project to which he did not devote sufficient time and effort to produce a truly satisfactory result. Yet, the size of the task alone serves to argue that Cooper was serious about its production: the *Continuation*, fifty close-set pages in length, is by far the longest of all the parts and versions previously published, and in the detail included (fullness of titles, extensiveness of listings of authors and editions, cross references, notes, etc.) it is the most complete. While the quality of Cooper's workmanship in preparing the entries is reasonably high, there are very serious collation problems, at least in the British Library copy, 270. b. 16. In this volume, the signatures have been incorrectly gathered near the middle of the text, i.e., D*2–15, D*16, D*19, D*20, D*17, D*18, D*21–D*26, and the resulting confusion can be overcome only though a laborious renumbering of the pages, which even then makes the *Continuation* very difficult to use. It is entirely possible that these serious errors in collation are responsible for the scarcity of copies of the *Continuation*: once Cooper became aware of how badly it had been assembled, he withdrew it from the market.[83]

---

[83] It should be noted, however, that the National Library of Medicine copy (Z/5522/C778c/1688) does not contain the collation problems present in the British Library copy. Like the British Library copy, it is not a part of the *Philosophical Epitaph*, but is bound between the second and third parts of the 1675 edition. The *Continuation* of the Library of Congress copy (Z/5522/.C77) follows Part Three.

That it was not reissued in a corrected version may have resulted from circumstances related to his death the following year; as we have seen, Plomer reports that this occurred before November, 1689, and that his business at the Pelican was taken over by his widow.

The importance of this *Continuation* results from the added comprehensiveness it gives to the *Catalogue*: by extending his book list thirteen years beyond the previous 1675 edition, Cooper has increased the total number of entries to nearly 430 and thus provided a much fuller inventory of books that had been published and were available for purchase in late Restoration England. Present space allows only a cursory glance at a few of the authors and titles now included in the expanded *Catalogue* or at a handful of the topics that might be profitably studied through investigation of works listed by Cooper.

Predictably, the revered authorities on magic, English and Continental, ancient and modern, are well represented in recent editions and translations: Agrippa, Albertus Magnus, Apollonius, Roger Bacon, Cyranus, Dee, Fludd, Porta, and Zoroaster (in Thomas Stanley's 1662 edition of *The History of the Chaldaick Philosophy*). Cabbala is represented in works by F. M. van Helmont and Henry More. As one might expect given the subject of Cooper's *Catalogue* and his own bibliographical interests, the range of alchemical works, both exoteric and esoteric, is enormous, and no certain lines can be drawn between these diverse titles and those of the emerging chemistry. Writings traditionally attributed to Hermes Trismegistus, both the *Pymander* and the *Asclepius*, are present in the recent translations of John Everard; and Isaac Casaubon's correct dating of these treatises notwithstanding, J[ohn] F[rench]'s (?) preface to the 1650 edition can still assert that "This Book may justly challenge the first place for antiquity, from all the Books in the World, being written some hundreds of yeers before *Moses* his time."[84] Arabian alchemical authorities are represented by Artephius and, more substantially, by the works of Geber (Jabir Ibn Haiyan) recently published in two editions of the Richard Russel translation. Among medieval alchemists, the only one of major importance not present is Arnold of Villanova; Flamel, Lull, Basil Valentine, Thomas Norton, George Ripley (extremely popular in the 16th and 17th centuries) and a host of others are present, and it is their influence along with that of Paracelsus and Boehme that is partially responsible for the interest in a wide range of alchemical thought in the seventeenth century. Cooper's *Catalogue* emphatically confirms what scholars of seventeenth century scientific and intellectual history have long contended: that the rise of the "New Science" was

---

[84] Hermes Trismegistus, *The Divine Pymander of Hermes Mercurius Trismegistus, in XVII Books. Translated . . . by that learned Divine Doctor Everard* (London, 1650), "To the Reader."

accompanied by, and was often indistinguishable from, vast quantities of writing on magic, alchemy, Hermeticism, astrology, and what is today considered theosophy.[85] Thus there is much Bacon in the *Catalogue* but more of Roger than Francis, and among works by single authors, those of Paracelsus (with seven main entries) and the Rosicrucians John Heydon and Thomas Vaughan (with nine and eleven titles respectively) are exceeded only by the twenty-five titles authored by Robert Boyle.

Along with alchemy and chemistry, medicine and related topics represent the largest single categories in Cooper's *Catalogue*. In addition to demonstrating the sheer bulk of writing on "physick" then available, Cooper provides an excellent means of following the several controversies that so vexed the theory and practice of medicine in the seventeenth century. The most important of these, the struggle between Galenist supporters of herbal cures and Paracelsian partisans of chemical preparations,[86] accounts for literally dozens of entries, which show that the latter were clearly winning the field even as they were being mercilessly skewered in the pages of their opponents' tracts. The Paracelsians included Oswald Croll, Joseph Duchesne (Quersetanus), Conrad Gesner, John French, George Starkey, Nicholas Culpeper, the Helmonts, and many others. Those whose titles suggest some willingness to compromise include George Castle, Moise Charas, Robert Godfrey and Walter Harris. The rough and tumble jurisdictional battles between the apothecaries and the College of Physicians are also amply represented, as are the strident claims of empirics like Richard Mathews and George Thomson. The earliest published book in the *Catalogue* is Hieronymous von Braunschweig's *Vertuose boke of Distyllacyon of the waters of all manner of Herbes* (1527), and one of the latest, and the only entry by a female writer, is Mrs. Mary Trye's *Medicatrix; or the Woman-Physician* (1675).

Related to the literature of physic and medical controversy and ex-

---

85 Recognition of the coexistence of these two traditions is not, of course, an argument that the earlier magical tradition was directly or indirectly responsible for the rise of the New Science. Debate continues concerning the "Yates' thesis," that man's will "to operate, stimulated by Renaissance magic, could pass into, and stimulate, the will to operate in genuine applied science," (see Frances A. Yates, *Giordano Bruno and the Hermetic Tradition* [Chicago: Univ. of Chicago Press, 1964], p. 150). At present, however, reaction against Yates' views is stronger than support for them. See, for example, the essays in the collection, *Occult and scientific mentalities in the Renaissance*, ed. Brian Vickers (Cambridge: Cambridge Univ. Press, 1984), and an earlier work instrumental in leading the reaction against Yeats, Robert S. Westman and J. E. McGuire, *Hermeticism and the Scientific Revolution*, Papers read at a Clark Library Seminar (Los Angeles: William Andrews Clark Memorial Library, 1977).

86 On the Paracelsian controversy in seventeenth-century England, see Allen G. Debus, *The English Paracelsians* (New York: Franklin Watts, 1966), especially chapters II, III and IV.

tremely popular throughout the century is the subject of baths and spas and the efficacy of mineral waters drawn from springs and fountains in England and on the Continent. Fully thirty-six titles by authors who today are largely forgotten (e.g. Thomas Guidott, Edward Jorden, William Simpson) attest to the popularity of resorts and salubrious waters long before "Beau" Nash and the Pump Room at Bath.

It has recently been argued that with the advent of printing, the genre of "books of secrets" played an important role, especially in Italy, in introducing the literate populace to science and technology and the necessity for first-hand investigation of Nature.[87] Written not by leading scientific thinkers but by "'middle-level' intelligentsia composed of professional writers, physicians, and nonacademics,"[88] these books traded in the experientially-derived wisdom of craft and trade, household and healing, rather than in the arcana of magic and alchemy. That England, too, provided a large audience for these "how-to" propaedeutics to technology and science is evident from Cooper's *Catalogue*. For example, two of Italy's leading producers of books of secrets, Alessio Piemontese, i.e., Girolamo Ruscelli (no. 6) and Leonardo Fioravanti (nos. 141–44) are represented by many translated editions dating back as early as 1559. These were joined by such English contributions to the genre as John Partridge's *Treasury of commodious conceits* (1591; no. 288) and Leonard Mascall's *Profitable boke declaring dyvers approoved remedies* (1596, 1605; no. 250). The probably supposititious *Secrets of Albertus Magnus* (no. 5) went through four editions from 1617–50.

Lest one is left with the view that Cooper's *Catalogue* presents only a view of an age in which learning is, on the one hand, practical and popular or, on the other, "vain, delicate, and contentious" (to use Bacon's words), it is equally apparent that the clarion calls of the *buccinator novi temporis* were not only heard but often anticipated. The titles of Bacon's *History Naturall and Experimentall* (no. 14) and his *Sylva Sylvarum, or a Natural History . . . whereunto is added Articles of Enquiry* (no. 15) reflect the dawning empirical and inductive emphasis, and in entries for works published during the next half century the cause of science and experimental method is increasingly manifest. We see this not only in Robert Boyle's many treatises touching nearly every aspect of natural philosophy, or in Henry Oldenburg's publication of the *Transactions of the Royal Society* (1665; no. 277), or in Thomas Sprat's *History* (no. 351). The growing attention given to experimentation and to "matter" rather than "words" is present in Henry Power's *Experimental Philoso-*

---

87 William L. Eamon, "Science and Popular Culture in Sixteenth Century Italy: The 'Professors of Secrets' and Their Books," *Sixteenth Century Journal*, XVI (1985), 471–85.

88 Eamon, p. 472.

phy, in three books, containing new experiments—microscopical, mercurial, magnetical—with some deductions and probable hypotheses raised from them (1664; no. 308) as well as in Robert Hooke's book on optics and magnification of the following year (no. 205). It is also present in the botanical works of Nehemiah Grew, Sir Kenelm Digby, and Hugh Platt; in William Borough's *Discours of . . . the Cumpas* (1596; no. 44); in those on mining by Thomas Houghton and Gabriel Plattes, and in the treatises on surgery by Daniel Sennertus.

If Cooper's *Catalogue* teaches us little that seems fundamentally new about the relationships between occultism and science in Renaissance and seventeenth-century England, it at least provides us with a treasure house of now forgotten material with which to continue our investigations of these "divided and distinguished" (or more accurately, I think, *connected and complementary*) worlds. Cooper is, after all, very much a part of the age of Sir Thomas Browne, and he emphasizes this "amphibiousness" in the Bookseller's address which prefaces the *Second Part* of the 1675 edition of the *Catalogue*. Seemingly uncomfortable over the inclusion of many "occult" titles in this listing of "chymicall" books, he asks the "Courteous Reader" to "Be pleased to understand, that some (*small Number*) of these Books in this Catalogue cannot absolutely be called Chymical, but have a very near affinity thereunto, the knowledge of natural Philosophy being an Introduction to supernatural things."[89]

A Note on this Edition

This edition of the *Catalogue of Chymicall Books* verifies and combines in a single alphabetical listing the versions of Cooper's work that appeared in 1673, 1675, and 1688. (The entries that appear in the 1688 *Continuation* are identified with an asterisk; because the emphasis in Cooper's work is on books and because other complete indices are available, I have omitted the third part of the 1675 edition which collects items on chemistry that were published in the *Philosophical Transactions* of the Royal Society during the first ten years.)

As sources of the 1673 and 1675 information, I have consulted editions of the *Philosophical Epitaph* and the *Catalogue* in the Library of Congress, the Henry E. Huntington Library, the Bodleian Library, and the British Library. My base text for the 1688 *Continuation* is British Library, 270. b. 16 and a microfilm copy supplied by the National Library of Medicine at Bethesda, Maryland (Z/5522/C778c/1688). This copy

89 "The Bookseller to the Reader," before sig. A.

does not contain the collation problems present in the British Library volume.

The verification process consisted, first, of searching for each of Cooper's authors and titles in the latest published volumes of the *Short-title Catalogue* or Wing to confirm their existence, and of recording identification numbers and other information which Cooper either does not supply, e.g. names of publishers, or supplies inconsistently, e.g. places of publication and volume sizes. In the cases of inconsistencies, discrepancies have been bracketed: e.g. "fol. [4°, Wing, BL]" indicates that Cooper reports the size of the volume as folio; Wing and the *British Library General Catalogue* record quarto; or "1575 [1579, STC]" indicates that Cooper reports the publication date as 1575, the *Short-title Catalogue* as 1579.

In addition to providing STC or Wing numbers, I have supplied for as many volumes as possible, a shelf-mark or call number from one major research library (usually the British Library) to confirm the actual existence of a given volume and to assist in its identification and locating. I have physically inspected many but not all of the titles and editions included in the *Catalogue*. Since my primary purpose has been to provide an accurate, useful, and usable edition of the three versions of the *Catalogue* that Cooper published, and since my concern is not with the physical characteristics of particular copies of books listed in them, I have not included descriptive bibliographical information pertaining to these works. In cases where I have not actually seen a volume, I have, wherever possible, derived its title from the *British Library General Catalogue* or from Cooper's own ample descriptions. In recording titles, I have been guided more by the principle of amplitude than parsimony; since many of the works are little known, I have attempted to reveal the contents of a book as fully as possible, particularly in the many instances where multiple authors and treatises are included in a single volume. In this respect I have followed the example of Cooper. In transcribing titles, I have adhered as closely as possible to the spellings and conventions of capitalization found on title pages, in *The British Library General Catalogue*, or in Cooper's own descriptions. Bracketed notes at the ends of entries contain additional bibliographical information, special notes provided by Cooper, or notices of his errors. The material included in Appendices I and II is derived from sources other than the *Catalogue*, and related information is included in the appendices themselves.

Abbreviations and Symbols

BL	*British Library* (or *British Museum*) *General Catalogue of Printed Books*
Bodleian	Bodleian Library, Oxford
CMCA	Samuel Hartlib, *Chymical, Medicinal and Chyrurgical Addresses, made to Samuel Hartlib, Esquire, in nine Treatises* (London, 1655)
Duveen	*Bibliotheca Alchemica et Chemica: An Annotated Catalogue of Printed Books on Alchemy, Chemistry and Cognate Subjects in the Library of Denis I. Duveen* (1949; rpt. London: Dawsons, 1965)
Ferguson	*Bibliotheca Chemica: A Bibliography of Books on Alchemy, Chemistry and Pharmaceutics.* 2 vols. (1906; rpt. London: Verschoyle, 1954)
Folger	*Catalog of Printed Books of the Folger Shakespeare Library*
Maunsell	Andrew Maunsell, *The Seconde parte of the Catalogue of English printed Bookes . . . Gathered into Alphabet, and such methode as it is, by Andrew Maunsell Bookseller* (London, 1595)
NUC	*National Union Catalogue*
pseud.	pseudonym
STC	Katharine F. Pantzer, W. A. Jackson, and F. S. Ferguson, *A Short-Title Catalogue of Books Printed in England, Scotland, & Ireland and of English Books Printed Abroad 1475–1640*, 2nd ed., revised and enlarged. 2 vols. (London: Bibliographical Society, 1976, 1986)

TC	*Term Catalogues, 1668–1709*, ed. Edward Arber, 3 vols. (London, 1903)
TCB	Elias Ashmole, *Theatrum Chemicum Britannicum* (London, 1652)
Thomason	Thomason Collection, British Library, London
UCLA	University Research Library, University of California, Los Angeles
Wellcome	Wellcome Institute for the History of Medicine, London
Williams	Dr. Williams's Library, London
Wing	Donald Wing, *Short-Title Catalogue of Books Printed in England, Scotland, Ireland, Wales, and British America and of English Books Printed in Other Countries 1641–1700*, 2nd ed., revised and enlarged. 3 vols. (New York: Modern Language Association, 1951, 1972, 1982)
*	Entry first appears in William Cooper, *The Continuation or Appendix to the Second Part of the Catalogue of Chymical Books, printed with The Philosophical Epitaph, being either Corrections, Rectifications, or Additions to the Editions of the old Books; and a Continuation of such New Books as have been Printed since the Year 1675. to this present Year 1688* (London, 1688)

Bibliography

Besant, Walter. *London City.* London: Adam & Charles Black, 1910.

__________*London in the Time of the Stuarts.* London: Adam & Charles Black, 1908.

Breger, Herbert. "Elias Artista—A Precursor of the Messiah in Natural Science." In *Nineteen Eighty-four: Science Between Utopia and Dystopia.* Ed. Everett Mendelsohn and Helga Nowotny. Dordrecht, Holland: D. Reidel, 1984, pp. 49–72.

The Complete Peerage, or a History of the House of Lords and all its Members from the Earliest Times. By G. E. C., rev. Vicary Gibbs, ed. H. A. Doubleday and Howard de Walden. 13 vols. London: St. Catherine Press, 1910–40.

Cunningham, George H. *London: Being a Comprehensive Survey of the History, Tradition & Historical Associations of Buildings and Monuments.* London: J. M. Dent, 1927.

Debus, Allen G. *The English Paracelsians.* New York: Franklin Watts, 1966.

Dobbs, Betty Jo Teeter. *The Foundations of Newton's Alchemy or "The Hunting of the Greene Lyon."* Cambridge: Cambridge Univ. Press, 1975.

Duveen, Denis I. *Bibliotheca Alchemica et Chemica: An Annotated Catalogue of Printed Books on Alchemy, Chemistry and Cognate Subjects in the Library of Denis I. Duveen.* 1949; rpt. London: Dawson's, 1965.

Eamon, William L. "Science and Popular Culture in Sixteenth Century Italy: The 'Professors of Secrets' and their Books." *Sixteenth Century Journal,* 16 (1985), 471–85.

Ferguson, John. *Bibliotheca Chemica: A Bibliography of Books on Alchemy, Chemistry and Pharmaceutics.* 2 vols. 1906; rpt. London: Derek Verschoyle, 1954.

__________"Some English Alchemical Books." *Journal of the Alchemical Society,"* 2 (1913), 2–16.

Ficino, Marsilio. *Commentary on Plato's Symposium on Love*. Ed. and trans. Sears Jayne. Dallas, Texas: Spring Publications, 1985.

Heninger, S. K., Jr. *The Cosmographical Glass: Renaissance Diagrams of the Universe*. San Marino, California: Huntington Library, 1977.

———*Touches of Sweet Harmony: Pythagorean Cosmology and Renaissance Poetics*. San Marino, California: Huntington Library, 1974.

Index to the Stationers' Register, 1640–1708. Ed. William P. Williams. La Jolla, California: Laurence McGilvery, 1980.

Johnson, Alfred Forbes. *A Catalogue of Engraved and Etched English Title-Pages*. Oxford: Oxford Univ. Press, 1934.

Josten, C. H., ed. *Elias Ashmole (1617–1692): His Autobiographical and Historical Notes, his Correspondence, and other Contemporary Sources Relating to his Life and Work*. 5 vols. Oxford: Clarendon Press, 1966.

Jung, C. G. *Psychology and Alchemy*. Trans. R. F. C. Hull. 2nd ed. Princeton: Princeton Univ. Press, 1953.

Lawler, John. *Book Auctions in England in the Seventeenth Century (1676–1700)*. 1898; rpt. Detroit: Gale Research, 1968.

Linden, Stanton J. "Alchemy and Eschatology in Seventeenth-Century Poetry." *Ambix*, 31 (November, 1984), 102–24.

———"Robert Burton's Sigil and John Dee." *Notes and Queries*, n.s. 29 (October, 1982), 415–16.

Macphail, Ian, comp. *Alchemy and the Occult: A Catalogue of Books and Manuscripts from the Collection of Paul and Mary Mellon Given to Yale University*. 2 vols. New Haven: Yale Univ. Library, 1968.

Maunsell, Andrew. *The First Part of the Catalogue of English printed Bookes . . . gathered into Alphabet, and such Method as it is, by Andrew Maunsell*. London, 1595.

———*The Seconde parte of the Catalogue of English printed Bookes . . . Gathered into Alphabet, and such methode as it is, by Andrew Maunsell*. London, 1595.

Morley, Frank. *Literary Britain: A Reader's Guide to Its Writers and Landmarks.* New York: Harper and Row, 1980.

Pantzer, Katherine F., W. A. Jackson, and F. S. Ferguson. *A Short-Title Catalogue of Books Printed in England, Scotland, & Ireland and of English Books Printed Abroad 1475–1640.* 2 vols. 2nd ed., rev. London: Bibliographical Society, 1976, 1986.

Plomer, Henry R. *A Dictionary of the Booksellers and Printers who were at Work in England, Scotland and Ireland from 1641 to 1667.* 1907; rpt. Oxford: Bibliographical Society, 1968.

_________*A Dictionary of the Printers and Booksellers who were at Work in England, Scotland and Ireland from 1668–1725.* 1922; rpt. Oxford: Bibliographical Society, 1968.

Plotinus. *The Enneads.* Ed. and trans. Stephen MacKenna. 2nd ed., rev. 1917–30; rpt. London: Faber and Faber, 1956.

Pritchard, Alan, comp. *Alchemy: A Bibliography of English-language Writings.* London: Routledge & Kegan Paul, 1980.

The Term Catalogues, 1668–1709 A.D. Ed. Edward Arber. 3 vols. London: privately printed, 1903–06.

A Transcript of the Registers of the Worshipful Company of Stationers; from 1640–1708 A.D. Ed. G. E. B. Eyre. 3 vols. 1913–14; rpt. Gloucester, Mass.: Peter Smith, 1967.

Vickers, Brian, ed. *Occult and Scientific Mentalities in the Renaissance.* Cambridge: Cambridge Univ. Press, 1984.

Waite, Arthur E. *Alchemists through the Ages.* 1888; rpt. [of *Lives of Alchemystical Philosophers*] Blauvelt, NY: Rudolf Steiner, 1970.

Webster, Charles. *The Great Instauration: Science, Medicine and Reform, 1626–1660.* New York: Holmes & Meier, 1975.

Westman, Robert S. and J. E. McGuire. *Hermeticism and the Scientific Revolution.* Papers read at a Clark Library Seminar, March 9, 1974. Los Angeles: William Andrews Clark Memorial Library, 1977.

Wilkinson, Ronald S. "The Alchemical Library of John Winthrop, Jr. (1606–

1676) and his Descendants in Colonial America." *Ambix*, 11 (February, 1963), 33–51, and 13 (October, 1966), 139–86.

_________"Further Thoughts on the Identity of 'Eirenaeus Philalethes.'" *Ambix*, 19 (November, 1972), 204–08.

_________"George Starkey, Physician and Alchemist." *Ambix*, 11 (October, 1963), 121–52.

_________"The Problem of the Identity of Eirenaeus Philalethes." *Ambix*, 12 (February, 1964), 24–43.

_________"Some Bibliographical Puzzles Concerning George Starkey." *Ambix*, 20 (November, 1973), 235–44.

Willard, Thomas S. "Alchemy and the Bible." In *Centre and Labyrinth: Essays in Honour of Northrup Frye*. Ed. Eleanor Cook, et al. Toronto: Univ. of Toronto Press, 1982, pp. 115–27.

Wing, Donald. *Short-Title Catalogue of Books Printed in England, Scotland, Ireland, Wales, and British America and of English Books Printed in Other Countries 1641–1700.* 3 vols. 2nd ed., rev. New York: Modern Language Association, 1951, 1972, 1982.

Yates, Frances A. *Giordano Bruno and the Hermetic Tradition.* Chicago: Univ. of Chicago Press, 1964.

A Catalogue of Chymicall Books

ABANO, Petrus de
Heptameron; or, Magical Elements (see AGRIPPA, *Fourth Book*)

1 **ACTON, George**
Letter in answer to certain quaeries and objections made by a Galenist against the theorie and practice of chymical physick, wherein the right Method of curing Diseases is demonstrated[,] the possibility of a Universal Medicine evinced, and Chymical Physick vindicated. Pp. 14.
London, William Godbid for Walter Kettleby
1670
4°
Wing A449; BL 551. a. 9. (3.)

2 **ACTON, George**
Physical Reflections upon a Letter written by J. Denis . . . to Monsieur de Montmor . . . concerning a new way of curing sundry diseases by Transfusion of Blood. Pp. 10.
London, T. R. for J. Martyn
1668
4°
Wing A450; BL 783. g. 23. (14.)

3 **AGRIPPA, Henricus Cornelius**
Henry Cornelius Agrippa his *Fourth Book of Occult Philosophy, Of Geomancy and Magical Ceremonies; to which is added Pet. de Abano his Heptameron or Magical Elements; Also Geo. Pictorius Villinganus Isagoge, or an Introductory Discourse of the Nature of such Spirits as are exercised in the sublunary bounds; their original, names, offices, illusions, powers, prophesies, miracles, and how they may be expelled and driven away; Likewise Gerh. Cremoninsis' of Astronomical Geomancy; and last of all Arbatel of the Magick of the Ancients, or the spiritual Wisdom of the Ancients, as well wise men of the people of God, as Magi of the Gentiles, &c., translated into English by Robert Turner.* [BL lists as Supposititious.] Pp. 217.
London, printed by J. C. for John Harrison
1655
4°
Wing A785; BL E. 833. (1.)

4 **AGRIPPA, Henricus Cornelius**
Three Books of Occult Philosophy, or of Magick . . . Translated . . . by J. F. [John French]. Pp. 583.
London, R. W. for Gregory Moule

1651
4°
Wing A789; BL 1471. g. 5.

[**ALAZONOMASTIX PHILALETHES**] (see MORE, Henry)

5 **ALBERTUS MAGNUS**, Bishop of Ratisbon
The Secrets of Albertus Magnus. Of the Vertues of Hearbes, Stones and certaine Beasts. Whereunto is newly added, a short discourse of the seven planets . . . Also a booke of the same author of the marvellous things of the world. [BL lists as Supposititious.]
London
*1617: W. Jaggard, 8° STC 265; BL C. 54. a. 23.
*1632: T. Cotes, sold by A. Kembe in Southwarke, 8° STC 266.5
1637: T. Cotes: sold by Andrew Kembe, 8° STC 267; BL Mic. A. 586. (2.)
*1650: by R. Cotes, to be sold by Fulke Clifton, 8° Wing A875I

6 **ALESSIO, Piemontese**, pseud. [**RUSCELLI, Girolamo**]
The Secretes of . . . Alexis of Piemount. Containyng excellente remedies against divers diseases, woundes, and other accidents . . . Translated out of French . . . by W. Warde. 4 pt.
London
*1559: [pt. 1], H. Sutton, 4° STC 295; BL C. 31. d. 11. [?]
1562: 3 pts., R. Hall for N. Englande, 4° STC 296; BL C. 113. b. 9.
*1563: [2nd. pt.], R. Hall for N. Englande, 4° STC 301; BL C. 31. d. 11. [?]
1568: 4 pts., N. Bynneman for J. Wight, 4° STC 297; BL 236. k. 19.
*1578: [3rd. pt.], T. Dawson for J. Wight, 4° STC 307; BL 541. b. 19.
1580: 4 pts., J. Kyngston for J. Wight, 4° STC 298; BL 541. b. 19.
1595: 4 pts., P. Short for T. Wight, 4° STC 312
*1614–5: [4th. pt.], W. Stansby for R. Meighen and T. Jones, 4° STC 311; BL 1508/1679 or 1608/519.

ALPHONSUS, King of Portugal (see P., H.)

7 **ALSTON, Sir Edward**
A Paper delivered in, To the honorable committee for Bathes and Bath-stoves, By Dr. Alston, Dr. Hamens, Dr. Bates, and Dr. Micklethwait, in the name of the President and Colledge of Physitians against Baths and Bath-stoves; with an Answer thereunto by Dr. Chamberlen. (see CHAMBERLEN, Peter, *Vindication* and *Public Bathes purged*)

London
1648
4°
Wing A2928; BL 551. b. 44.

ANDREWES, Abraham (see ASHMOLE, *TCB*)

ANON
Description of the Stone (see ASHMOLE, *TCB*)

ANON
Experience and Philosophy (see ASHMOLE, *TCB*)

ANON
Fragments of the Philosophers (see ASHMOLE, *TCB*)

ANON
Hermes Bird (see ASHMOLE, *TCB*)

ANON
The Hermits Tale (see ASHMOLE, *TCB*)

ANON
An Invitation to a free and generous communication of Secrets (see HARTLIB, *CMCA*)

ANON
Liber Patris Sapientae (see ASHMOLE, *TCB*)

ANON
A Profitable discourse Composed by divers Grocers against the bad Garbelling of Spices (see *Profitable discourse*)

ANON
A Prognostication of what shall happen to Physitians, Chirurgians, Apothecaries, and their dependents; with Chymists and Miners (see HARTLIB, *CMCA*)

ANON
The Standing of the Glass for the Time of Putrifaction and Congelation of the Medicine (see ASHMOLE, *TCB*)

ANON
Whether the Urim and Thummim were given in the Mount, or perfected by Art; and that it is of like pure substance with the White and Red Elixirs (see HARTLIB, *CMCA*)

8 ANTHONIE, Dr. Francis
The Apologie, or defence of a verity heretofore published concerning a medicine called Aurum Potabile. Pp. 126.
London, John Legatt
1616
4°
STC 666; BL 1142. h. 28.

9 **Anthropologie abstracted: or, the Idea of humane nature reflected in briefe philosophicall, and anatomicall collections.* Pp. 201.
London, for Henry Herringman
1655
8°
Wing A3483; BL E. 1589. (2.)

10 APOLLONIUS, Tyaneus
Ars Notoria: the Notory Art of Solomon, shewing the cabalistical key of magical operations, the liberal sciences, divine revelation, and the art of memory. Whereunto is added an Astrological Catechism, fully demonstrating the art of Judicial Astrology . . . Written originally in Latine [by Apollonius, Leovitius, and others. Collected] *and now Englished by R. Turner.*
London, by J. Cottrel, to be sold by Martha Harison
1657
8°
Wing A3533; BL E. 1655. (2.)

ARBATEL
Of the Magick of the Ancients (see AGRIPPA, *Fourth Book*)

ARTEPHIUS
His Key of the Greater Wisdom (see FLAMEL, Nicholas)

ARTEPHIUS
Secret Book of the Blessed Stone, called the Philosophers (see FLAMEL, Nicholas, *Exposition of the hierogliphical figures*)

ASHMOLE, Elias, translator
Fasciculus chemicus (see DEE, Arthur)

11 ASHMOLE, Elias
Theatrum Chemicum Britannicum. Containing Severall Poeticall Pieces of our Famous English Philosophers, who have written the Hermetique Mysteries in their owne Ancient Language. Faithfully Collected into one Volume, with Annotations thereon. Pp. 486 [and Tables].
London, J. Grismond for Nath. Brooke
1652
4° [8°, Wing]
Wing A3987; BL 239. k. 6.

Contents:
1. **NORTON, Thomas**, *Ordinall of Alchemie*, p. 1.
2. **RIPLEY, George**, *Compound of Alchemie*, p. 107.
3. **ANON.**, *Pater Sapientiae*, p. 194.
4. **ANON.**, *Hermes's Bird*, p. 211.
5. **CHAUCER, Geoffry**, *Chanon's Yeoman's Tale*, p. 227.
6. **DASTIN, John**, *Dastin's Dreame*, p. 257.
7. **PEARCE** (the Black Monke), *upon the Elixir*, p. 269.
8. **CARPENTER, Richard**, *Worke*, p. 275.
9. **ANDREWES, Abraham**, *Hunting of the Greene Lyon*, p. 278.
10. **CHARNOCK, Thomas**, *Breviary of Naturall Philosophy*, p. 291.
11. **CHARNOCK, Thomas**, *Aenigmaes*, p. 303.
12. **BLOOMEFIELD, William**, *Bloomfields Blossomes*, p. 305.
13. **KELLEY, Sir Edward**, *Worke*, p. 324.
14. **KELLEY, Sir Edward**, *to G.S. Gent.*, p. 332.
15. **DEE, Doctor John**, *Testament*, p. 334.
16. **ROBINSON, Thomas**, *Of the Philosophers Stone*, p. 335.
17. **ANON.**, *Experience and Philosophy*, p. 336.
18. **W. B.**, *The Magistery*, p. 342.
19. *Anonymi, or severall Workes of unknowne Authors*, pp. 334 & 404.
20. **GOWER, John**, *Upon the Philosophers Stone*, p. 368.
21. **RIPLEY, George**, *Vision*, p. 374.
22. *Verses belonging to Ripley's Scrowle*, p. 375.
23. *Mystery of Alchemists*, p. 380.
24. **RIPLEY, George**, *Preface to the Medulla*, p. 389.
25. **RIPLEY, George**, *A short Worke*, p. 393.
26. **LYDGATE, John**, *Secreta Secretorum*, p. 397.
27. **ANON.**, *Hermit's Tale*, p. 415.
28. **ANON.**, *Discription of the Stone*, p. 420.

29. **ANON.**, *The Standing of the Glasse*, &c., p. 421.
30. **REDMAN, W.**, *Aenigma Philosophicum*, p. 423.
31. *Fragments*, p. 424

12 ASHMOLE, Elias
The Way to Bliss. In three books, etc. Pp. 220.
London, John Grismond for Nath. Brooke
1658
4°
Wing A3988; BL E. 940. (3.)

B., D.
**Rare and excellent discourse of Minerals, Stones, Gums, Rosins, . . .*
(see PLATT, *Jewell House*, 1653 edition)

13 BACON, Francis
**Baconiana. Or Certain Genuine Remains of Sr. Francis Bacon . . . In arguments civil and moral, natural, medical, theological, and bibliographical; now the first time faithfully published. An account of these Remains, and all of his Lordship's other works, is given by the publisher, in a discourse by way of introduction* [signed: T. T., i.e., Thomas Tenison, Archbishop of Canterbury]. 2 pt. [With Dr. Meverell's Answers] (see MEVERELL)
London, J. D. for Richard Chiswell
1679
8°
Wing B269; BL 535. b. 1.

14 BACON, Francis
**History Naturall and Experimentall, of Life and Death; or Of the prolongation of life.*
London, John Haviland, for W. Lee & H. Mosely
1638
8° [12°, BL]
STC 1158; BL 535. a. 6.

15 BACON, Francis
Sylva Sylvarum, or a Natural History . . . whereunto is added Articles of Enquiry, touching Metals and Minerals (ninth edition). 5 pt.
London
1670: J. R., for William Lee, fol. Wing B331; BL 986. g. 14.
*1677: [reprint of tenth edition], by S. G. and B. Griffin for Thomas

Lee, fol. [8°, BL] Wing B333; BL 535. i. 12. (1.)

16 **BACON, Roger**
Art of Chymistry.
16°
[Probably an unlisted translation of Bacon's *Sanioris medicinae magistri D. Rogeri Baconis Angli, De arte chymiae scripta. Cui accesserunt opuscula alia ejusdem authoris.* Frankfurt, typus Joannis Saurii, 1603, reissued 1620. Pp. 408, NUC 0023161; BL 1036. a. 11.; UCLA QD 25/B13s]

17 **BACON, Roger**
**The Cure of Old Age, and Preservation of Youth. By Roger Bacon . . . Translated out of Latin* [i.e., from the abridgment published in 1590]; *with annotations and an account of his life and writings. By Richard Browne . . . Also a Physical Account of the tree of life by Edw. Madeira Arrais, translated likewise out of Latin by the same hand.* 2 pt.
London, for Tho. Flesher and Edward Evets
1683
8° [12°, BL]
Wing B372; BL 1039. d. 22.

18 **BACON, Roger**
Discovery of the Miracles of Art, Nature, and Magick. Faithfully translated out of Dr. Dees own copy, by T. M. and never before in English.
Pp. 51.
London, for Simon Miller
1659
12°
Wing B373; BL E. 1932. (1.)

BACON, Roger
Medicine or tincture of Antimony to preserve health, and cure the leprosie of Metals, to purifie and transmute them into the best of Gold (see VALENTINE, Basil, *Of Natural and Supernatural Things*)

19 **BACON, Roger**
Mirror of Alchimy, . . . or a most excellent and learned discourse of the admirable force and efficacie of Art and Nature; also the Smaragdine Table of Alchimy, by Hermes Trismegistus; with a briefe Commentary of Hortulanus upon the Smaragdine Table, with the books of the Secrets of Alchimie composed by Galid the son of Jazich, translated out of Heb.,

Arab., and Lat., into English. Pp. 84.
London, Thomas Creede for Richard Olive
1597
4°
STC 1182; BL C. 115 R. 11.

20 BACON, William
**A Key to* [Jean Baptiste van] *Helmont. Or, a short introduction to the better understanding of the theory and method of the most profound chymical physitians.* Pp. 34.
London, for John Starkey
1682
4°
Wing B 374; BL 1034. h. 35.

21 BADCOCK, William
**A touch-stone for gold and silver wares. Or, a manual for goldsmiths . . . Together with the several statutes now in force for regulating abuses committed in that craft.* Pp. 115.
London, printed for John Bellinger and Thomas Bassett
1677: 8° Wing B382; BL 1651/170.
1679: [second edition, to which is added: *The useful and easie tables of Mr. John Reynolds*] 8° Wing B381; BL 1651/1059.

22 BAILEY [BAILY], **Walter**
**A Briefe Discours of certain Bathes or medicinall Waters in the Countie of Warwick neere unto a village called Newnam Regis.* Pp. 35.
London, Ent. to J. Charlwood
1687 [1587]
8°
STC 1191; BL 1171. c. 40.

23 BAKER, George
**The Composition or Making of the most excellent and pretious Oyl called Oleum Magistrale. First published by the commaundement of the King of Spain, with the maner how to apply it particulerly . . . Also the third book of Galen of curing of pricks and wounds of Sinowes [or rather, of the composition of medicines]. A method for curing of wounds in the joynts . . . A breef gathering togither of certain errours which the common chirurgians dayly use . . . Gathered and translated by G. Baker.* Ff. 51.
London, J. Alde

1574
8°
STC 1209; BL C. 31. a. 17.

BAKER, George (see CLOWES, William and GESNER, Conrad)

24 BARBA, Alvaro Alonso
Two Books of the Art of Mettals; In the First is declared the manner of their Generation, and the concomitants of them. In the Second is taught the common way of refining Silver by Quick-silver; with some new Rules added for the better performance of the same: translated out of the Spanish into English by Edward Earl of Sandwich. 2 vols.
London
1670: For S. Mearne, 8° Wing B679, B681; BL 987. b. 25, 26.
1674: For S. Mearne, 8° Wing B680, B682; BL 446. a. 25. (3.)

25 BARLOW, William
**A Briefe Discovery of the idle Animadversions of Marke Ridley . . . upon a Treatise entitled, Magneticall Advertisements . . . two sheets.* Pp. 13.
London, E. Griffin for Timothy Barlow
1618
4°
STC 1443; BL 538. f. 43.

26 BARLOW, William
Magneticall Advertisements, or divers pertinent Observations and approved Experiments concerning the Nature and Properties of the Loadstone. [Partly compiled from William Gilbert's "De Magnete".] Pp. 86.
London, E. Griffin for Timothy Barlow
1616
4°
STC 1442; BL C. 112. b. 9.

27 BECHER, Johann Joachim
**Magnalia Naturae: or, the Philosophers-stone lately expos'd to public sight and sale. Being a true and exact account of the manner how Wenceslaus Seilerus . . . came by and made away with a very great quantity of powder of projection.* Pp. 31.
London, by Tho. Dawks. Sold also by La. Curtiss
1681 [1680, Wing, BL]

4°
Wing B1643; BL 1035. c. 42.

28 BEGUINES, Joseph
Tyrocinium Chymicum; or, Chymical Essays, acquired from the fountain of nature, and Manual Experience [Translated by Richard Russell]. Pp. 136.
London, for Thomas Passenger
1669
8°
Wing B1703; BL 1651/1187

29 BELLON, Peter
**The Irish Spaw; being a short discourse on mineral waters in general . . . And a brief account of the mineral waters at Chappel-Izod near Dublin.* Pp. 76.
Dublin, J. R. for M. Gunne, and Nat. Tarrant
1684
12°
Wing B1852; BL 1171. e. 20. (3.)

30 BELLON, Peter
**Potable Balsome of Life, being a Collection of the choicest Preservatives . . . reduced into such Essences and Tinctures by the scientifick Art of Chymie . . . which constitute a radical and balsamick Liquor.*
London, for the author
1675
4°
Wing B1856 [cancelled in second edition]; Wellcome 13204/A.

BERNARD, Earl of Trevisan (see TREVISAN)

31 BERTHOLDUS, Andrew
Of the Wonderfull and strange effect and vertues of a new Terra Sigillata lately found out in Germanie . . . With the right order of the applying and administering of it, etc. [Translated by B. G., i.e., Barnaby Googe?] Pp. 39.
London
1587: R. Robinson, for R. Watkins, 8° STC 1970; BL 546. b. 33.
1589: [not listed in STC or BL]

32 BIGGS, Noah
Mataeotechnia medicinae praxews. The Vanity of the Craft of Physick; or a New Dispensatory, wherein is dissected the errors, ignorance, impostures and supinities of the schools in their main pillars of purges, blood-letting, and Medicines of the Shops; with an humble motion for the reformation of the universities and the whole landscape of physick in the further persuit of Chymistry proposed to the Parliament of England. Pp. 232.
London, Edward Blackmore [Giles Calvert, Wing]
1651
4°
Wing B2888; BL 551. a. 41.

33 BLANKAART, Stephen
**A Physical Dictionary; in which all terms relating either to anatomy, chirurgery, pharmacy, or chymistry, are very accurately explain'd.* Pp. 302.
London, by J. D. for J. Gellibrand
1684
8°
Wing B3164A; BL 1488. aa. 19.

BLOOMEFIELD, William
Blossoms (see ASHMOLE, *TCB*)

34 BOATE, Gerard
Irelands Naturall History, &c. Being a true and ample description of its situation . . . of the Springs, Fountains, Brooks, Rivers, Lakes, Metals, Minerals, Free-stones, Marble, Sea-coale, and other subterraneals . . . Now published by Samuel Hartlib. Pp. 186.
London
1652: John Wright, 8° Wing B3372; BL 957. c. 27
*1657: [a reissue], for John Wright, 8° Wing B3373; BL 957. c. 28.

35 BOEHME, Jacob
Aurora. That is, the Day-spring, or dawning of the day in the orient, or the morning redness in the rising of the Sun. That is the root or mother of philosophie, astrologie, and theologie, from the true ground; Or a description of Nature as she was originally, how changed creaturally, how she stands at present, and will be at the end, [Translated by J. Sparrow]. Pp. 643.
London, John Streater for Giles Calvert

1656
4°
Wing B3397; BL E. 890.

36 BOEHME, Jacob
The Second Booke. Concerning the three Principles of the Divine Essence of the eternall, dark, light, and temporary World: shewing what the Soul, the Image and the Spirit of the Soul are, &c.; how all things have been, now are, and shall be at last, [Translated by J. Sparrow]. Pp. 396.
London, M. S. [Matthew Simmons] for H. Blunden
1651 [1648 Wing, BL]
4°
Wing B3417; BL 853. g. 3.

37 BOEHME, Jacob
Signatura Rerum, or the Signature of all things, shewing the Sign and Signification of the severall Forms and Shapes in the Creation, and what the beginning, ruin, and cure of everything is; (The 177 Theosophick Questions); it proceeds out of Eternity into Time; and again, out of Time into Eternity, and compriseth all Mysteries, [Translated by J. Ellistone]. Pp. 224.
London, Giles Calvert
1651
4°
Wing B3419; BL E. 633. (3.)

38 BOEHME, Jacob
The Third Booke of the Authour being the High and Deepe Searching out of the Threefold Life of Man through, or according to, the Three Principles . . . Englished by J. Sparrow. Pp. 288.
London, M. S. for H. Blunden
1650
4°
Wing B3422; BL 1478. e. 3.

39 BOLNEST, Edward
Aurora chymica: or, a Rational way of preparing animals, vegetables and minerals for physical use, and preservation of the life of Man, 3 pts. Pp. 146.
London, Thomas Ratcliffe and Nat. Thompson, for John Starkey
1672
8°

Wing B3497; BL 1035. b. 10.

40 **BOLNEST, Edward**
Medicina instaurata: or a Brief account of the true grounds and principles of the art of physick made by chymical operation; and the insufficiency of the vulgar way of preparing medicines. Also an epistolary discourse upon the whole, by the author of Medela medicinae (Mar. Nedham). Pp. 151.
London, John Starkey
1665
8° [12°, Wing, BL]
Wing B3498; BL 1035. a. 15. (1.)

41 **A Booke of Secrets: shewing divers waies to make and prepare all sorts of Inke, and Colours . . . Also to write with Gold and Silver . . . with many other profitable secrets . . . Translated out of Dutch . . . by W. P.* [i.e., W. Philip. A translation of "Ettliche Künste, auff mancherley weisz Dinten und allerhand Farben zu bereyten."] *Hereunto is annexed a little Treatise, intituled, Instructions for ordering of Wines . . . Written, first in Italian, and now newly translated . . . by W. P.*
London, Adam Islip for Edward White
1596
4°
STC 3355; BL C. 31. c. 21.

42 **BORDER, Daniel**
Πολυφαρμακος και Χυμιστης; *or, the English unparallel'd Physician and Chyrurgian; shewing the true use of all manner of plants and minerals. In which is explained, the whole art and secresy of physick and chirurgery in the extracts of spirits and quintessences of herbs and minerals, salts, oyls, balms, &c.* Pp. 144.
London, B. Alsop
1651
4°
Wing B3751; BL 1038. i. 35. (6.)

43 **B**[ORLASE], **E**[dmund]
Latham Spaw in Yorkshire [*Lancashire*, Wing, BL] *with some remarkable cases and cures effected by it.* Pp. 72.
London, for Robert Clavel
1670: 8° Wing B3769; BL 1171. e. 19.
1672: 8° Wing B3770

44 BOROUGH, William
A Discours of the Variation of the Cumpas, or Magneticall Needle. Wherin is mathematically shewed, the maner of the observation, effectes, and application therof, made by W. B. And is to be annexed to The Newe Attractive of R. N. [R. Norman].
London, E. Allde for Hugh Astley
1596
4° [8°, Wing, BL]
STC 3391; BL 51. c. 7. (2.)

45 BOSTOCK, Robert
The Difference betwene the Auncient Physicke, first taught by godly Forefathers; consisting in unitie, peace and concorde: and the latter Physicke proceeding from idolaters, ethnickes, and heathens, as Gallen and such other.
London, G. Robinson for R. Walley
1585
8°
STC 1064; BL 1038. a. 7.

46 BOULTON, Samuel
Medicina Magica Tamen Physica . . . Magical, but Natural Physick. Or a Methodical Tractate of Diastastical Physick. Containing the general Cures of all infirmities . . . with a Description of the most excellent Cordial of Gold. Pp. 195.
London
1656: by T. C. for Nath. Brooke, 8° Wing B3833A; BL E. 1678. (2.)
*1665: 8° Wing B3833B

47 BOYLE, Robert
**The Aerial Noctiluca: or, some new phoenomena, and a process of a factitious self-shining substance, etc.* Pp. 109.
London, by T. Snowden to be sold by Nath. Ranew
1680
8°
Wing B3925; BL 1034. g. 30.

48 BOYLE, Robert
Certain Physiological Essays. *The second edition. Wherein some of the tracts are enlarged and the work is increased by the addition . . . Physiological Essays touching Experimental Essays in general; two Essays concerning the unsuccessfulness of Experiments; Some Specimens of an*

attempt to make Chymical Experiments, useful to illustrate the notions of the corpuscular Philosophy; the History of Fluidity and Firmness; and lastly, of absolute rest in Bodies. 2 pts.
London, for Henry Herringman
1669
4°
Wing B3930; BL 431. a. 21.

49 BOYLE, Robert
A Continuation of New Experiments Physico-Mechanical touching the Spring and Weight of the Air and their effects; whereunto is annexed a short discourse of the Atmospheres of Consistent Bodies, with eight Scheams. Pp. 198.
Oxford, H. Hall for Ric. Davis
1669
4°
Wing B3934; BL 535. c. 7. (4.)

50 BOYLE, Robert
Essay about the Origine & Virtues of Gems. Wherein are propos'd and historically illustrated some conjectures about the consistence of the matter of precious stones, and the subjects wherein their chiefest virtues reside. Pp. 185.
London, printed by William Godbid; sold by Moses Pitt
1672
8°
Wing B3947; BL 990. b. 2.

51 BOYLE, Robert
Essays of the Strange Subtilty, Determinate Nature, and Great Efficacy of Effluviums. To which are annext New experiments to make fire and flame ponderable: together with a Discovery of the perviousness of glass. 4 pt.
London, Wm. Godbid for M. Pitt
1673
8°
Wing B3951; BL 535. b. 19.

52 BOYLE, Robert
**Experiments and Considerations about the Porosity of Bodies, in two essays.* Pp. 145.
London, for Samuel Smith
1684

8°
Wing B3966; BL 535. b. 22.

53 BOYLE, Robert
Experiments and Considerations touching Colours. First occasionally written, among some other Essays, to a friend; and now suffer'd to come abroad as the beginning of an experimental history of colours; with an account of some Observations about a Diamond that shines in the dark. Pp. 423.
London, for Henry Herringman
1664: 8° Wing B3967; BL 958. a. 23.
1670: 8° Wing B3968; BL 1490. m. 45.

54 BOYLE, Robert
**Experiments, Notes, &c. about the Mechanical Origine or Production of divers particular Qualities: among which is inserted a Discourse of the imperfection of the chymist's doctrine of qualities; together with some Reflections upon the hypothesis of alcali and acidium. Viz. of the Mechanical Origine of Heat and Cold; of the Mechanical production of Tastes; of the Mechanical production of Odours; Volatility; Fixtness; Electricity; Corrosiveness and Corrosibility; Magnetism, etc.* 11 pts.
London, E. Flesher for R. Davis, Oxford
1676 [separate title pages of all parts except "of Magnetism" are dated 1675]
8°
Wing B3977; B3969, B3971, B3973, B3974, B3975; BL C. 175. m. 2.

55 BOYLE, Robert
[An Historical Account] *of a Degradation of Gold, made by an anti-elixir: a strange chymical narative* [sic]. Pp. 17.
London, by T. N., for Henry Herringman
1678
4°
Wing B3984; BL 1035. c. 44.

56 BOYLE, Robert
Hydrostatical Paradoxes, made out by new experiments, for the most part physical and easie. Pp. 247.
Oxford, Wm. Hall for R. Davis
1666
8°
Wing B3985; BL 1651/710

57 BOYLE, Robert
**New Experiments and Observations made upon the Icy Noctiluca . . . To which is annexed a Chymical Paradox.* Pp. 150.
London, R. E. for B. Tooke
1681–2
8°
Wing B3995; BL 1034. g. 31.

58 BOYLE, Robert
New Experiments and Observations touching Cold, or an Experimental history of cold, begun. To which are added an Examen of Antiperistasis, and an Examen of Mr. Hobs's Doctrine about Cold . . . Whereunto is annexed An Account of Freezing, brought in to the Royal Society, by the learned Dr. C. Merret, a Fellow of it. Pp. 845.54.
London
1665: for John Crooke, 8° Wing B3996; BL C. 76. a. 23.
*1683: [second edition with an appendix, pp. 324. 20. 29.], for Richard Davis, 4° Wing B3997; BL 462. c. 28.

59 Boyle, Robert
New Experiments Physico-Mechanicall, touching the Spring of the Air, and its effects, made, for the most part, in a new pneumatical engine. Pp. 339.
Oxford, H. Hall for Tho. Robinson
1660
8°
Wing B3998; BL C. 112. aa. 4.

60 BOYLE, Robert
New Experiments Physico-Mechanical, touching the Spring of the Air and its effects, with Figures, the Second Edition; with a Defence of the Doctrine touching the spring and weight of the Air, against the Objections of Fran. Lynus; with an Examen of Mr. Hobbes his Dialogus Physicus de Natura Aeris; with an Appendix touching Mr. Hobbes's Doctrine of Fluidity and Firmness. 3 pt.
Oxford
1662: 3 pts. H. Hall for Tho. Robinson, 4° Wing B3999; BL 535. c. 7. (1–3.)
1669: 3 pts. H. Hall for Richard Davis, 4° Wing B3934; BL 535. c. 7. (4.)
*1682: [third ed.] Miles Flesher for R. Davis, 4° Wing B4000; BL 461. a. 27. (1–3.)

61 BOYLE, Robert
**Of the Reconcileableness of Specifick Medicines to the Corpuscular Philosophy. To which is annexed a discourse about the advantages of the use of simple medicines.* Pp. 225.
London, for Samuel Smith
1685
8°
Wing B4013; BL 547. c. 11.

62 BOYLE, Robert
Origine of Formes and Qualities, according to the Corpuscular Philosophy; Illustrated by Considerations and Experiments, written by way of notes upon an Essay about Nitre. Pp. 433.
Oxford, H. Hall for R. Davis
1666: 8° Wing B4014; BL 1032. b. 29
1667: 8° Wing B4015; BL 535. b. 16.

63 BOYLE, Robert
The Sceptical Chymist: or Chymico-physical Doubts & Paradoxes, touching the spagyrist's principles commonly called hypostatical, as they are wont to be propos'd and defended by the generality of alchymists. Whereunto is premis'd part of another discourse relating to the same subject. Pp. 436.
London, J. Cadwell for J. Crooke
1661
8°
Wing B4021; BL C. 112. aa. 5.

64 BOYLE, Robert
**Short Memoirs for the Natural Experimental History of Mineral Waters. Addressed by way of a letter to a friend.* Pp. 112.
London, for Samuel Smith
1684–[5]
Wing B4023; BL 1171. d. 13.

65 BOYLE, Robert
Some Considerations touching the Usefulness of Experimental Naturall Philosophy; propos'd in familiar discourses to a friend, by way of invitation to the study of it . . . the first part of its usefulness in reference to the mind of man; the second part of its usefulness to Physick, 2 tom.
[1st edition]
Oxford, H. Hall for R. Davis
1663, 1664

4°
Wing B4029, B4030; BL 535. c. 14.

66 BOYLE, Robert
Some Considerations touching the Usefulness of Experimental Naturall Philosophy: Tome the Second, or of the Second Part the second Section of its usefulness to the Empire of man over inferiour Creatures; of the usefulness of Mathematicks to natural Philosophy; of the usefulness of Mechanical Disciplines to natural Philosophy, that the good of Mankind may be much encreased by Naturallists insight into Trades; of doing by Physical knowledge what is wont to require manual skill; and lastly, of mens great Ignorance of the uses of natural things, 2 tom. [2nd edition]
Oxford, H. Hall for R. Davis
1671
4°
Wing B4031; BL 535. c. 9.

67 BOYLE, Robert
Tracts, consisting of Observations about the Saltness of the Sea: an Account of a Statical Hygroscope and its uses; together with an appendix about the force of the air's moisture: a fragment about the natural and preternatural state of bodies . . . To all which is premis'd a sceptical dialogue about the positive or privative nature of cold; with some experiments of Mr. Boyl's referr'd to in that discourse. 9 pts.
London, E. Flesher for R. Davis in Oxford
1673 & 1674 [1674, Wing, BL]
8°
Wing B4053; BL 535. b. 20.

68 BOYLE, Robert
Tracts: containing I. Suspicions about some hidden qualities of the air; with an appendix touching celestial magnets . . . II. Animadversions upon Mr. Hobbes's Problemata in Vacuo. III. A Discourse of the cause of attraction by suction. 6 pts.
London, printed by W[illiam] G[odbid]; sold by M. Pitt
8°
1674: Wing B4054; BL 535. b. 21.
1675: Wing B4054A

69 BOYLE, Robert
Tracts written by the Honourable Robert Boyle. About The cosmicall qualities of things. Cosmicall suspitions. The temperature of the subterraneall regions. The temperature of the submarine regions. The bottom

of the sea. To which is prefixt, an Introduction to the history of particular qualities. 6 pt.
Oxford, W. H. for Ric. Davis
1671
8°
Wing B4057; BL Eve. a. 84.

70 BOYLE, Robert
Tracts written by the Honourable Robert Boyle, containing new experiments touching the relation betwixt flame and air. And about explosions. An hydrostatical discourse occasion'd by some objections of Dr. Henry More; with an Experiment about the way of weighing water in water . . . Of positive or relative Levity of bodies under water. Of the Air's spring on bodies under water, and about the differing pressure of heavie Solids and Fluids. 6 pt.
London, printed for R. Davis, Oxford
1672
8°
Wing B4060; BL 535. b. 18.

71 BOYLE, Robert
**Tracts written by the Honourable Robert Boyle, Fellow of the R[oyal] Society. Of a discovery of the admirable rarefaction of the air. New observations about the duration of the spring of the air. New experiments touching the condensation of the air by meer cold; and its compression without mechanical engins. The admirably differing extension of the same quantity of air rarefied and compressed.* Pp. 28.
London, T. N. for Henry Herringman
1671
4°
Wing B4059; BL 535. c. 7. (5.)

72 BRAUNSCHWEIG, Hieronymous von
The vertuose boke of Distyllacyon of the waters of all maner of Herbes, with the Figures of the Stillatories . . . translated out of Duyche.
London, Laurens Andrewe
1527 (April 18)
fol.
STC 13435; BL 448. g. 1.

73 *Brevis Responsio, being a Short Replication in answer to the late Discourse of Dan. Cox and his Complices, against the Society of Apothecaries, in English Verse.* (see COXE, Thomas or D[aniel])

London
1669
4°

[Additional information unavailable.]

A Brief Preparation of the Philosophers Stone (see HOUPREGHT, *Aurifontina Chymica*)

74 BROWN, Thomas
Nature's Cabinet unlock'd. Wherein is discovered the natural causes of metals, stones, precious earths, juices, humours, and spirits, the nature of plants in general; the affections, parts and kinds in particular, &c . . . together with a description of the individual parts and species of all animate bodies . . . by Tho. Brown D. of Physick. Pp. 331.
London, for Edward Farnham
1657
12°
Wing B5065; BL E. 1687.

[BL lists as "supposititious" under Sir Thomas Browne.]

75 BROWNE, Edward
**An Account of Several Travels through a great part of Germany: in four Journeys . . . Illustrated with sculptures.* Pp. 179.
London, for Benjamin Tooke
1677
4°
Wing B5109; BL 1048. i. 12.

76 BROWNE, Edward
A Brief Account of Some Travels, . . . As also some observations on the gold, silver, copper, quick-silver, mines, . . . in those parts, etc. Pp. 144.
London
1673: T. R. for Benj. Tooke, 4° Wing B5110; BL 982. d. 14.
*1687: [second edition reissued], for B. Tooke, sold by Tho. Sawbridge, fol. Wing B5112; BL 567. k. 1.

77 BUNWORTH, Richard
Ομοτροπια Naturae. A physical discourse, exhibiting the cure of diseases by signature, wherein is demonstrated a similitude of substance between Actives and Passives, and all operation of contrariety denied by demonstration of Poyson: together with a Tract called Man in Paradise,

or a Philosophical Discourse of the Souls Prerogative, &c. 2 pt.
London, Printed by J[ames] C[ottrel] & to be sold by Jer. Hirons
1655 [1656, Wing, BL]
12°
Wing B5478; BL 1038. a. 13.

77A BUNWORTH, Richard
**Rational Discourse of Coral and Steel grounded upon Experience.*
1656
12°

[This title is probably included in the preceding entry or in Bunworth's *Two Excellent Discourses*, London, for S. Waterson, 1657, 12°, Wing 5478A.]

BURGGRAV, Johannes E.
Introduction to Astral Philosophy (see BURGGRAV, *Method of chymical philosophie and physick*)

78 BURGGRAV, Johannes E.
The Method of chymical philosophie and physick. Being a brief Introduction to the one, and a true discovery of the other.
London, by J. G. for Nath. Brook
1664
8°
Wing M1943; BL 1651/1227

[Cooper says "published by Humph. Blunden Bookseller, but his Name is not printed to the book[,] printed by his servant, as was Major Brooks."]

79 BUSHELL, Thomas
An Extract by Mr. Bushell of his late Abridgement of the Lord Chancellor Bacons Philosophical Theory in Mineral Prosecutions.
London, for Tho. Leach
1660
4°
Wing B297; BL C. 27. f. 1. (6.)

80 BYFIELD, Timothy
**The Artificial Spaw, or Mineral-waters to drink: imitating the German spaw-water in its delightful and medicinal operations on Humane Bodies, &c.* Pp. 70.
London, James Rawlins for the Author

1684
12°
Wing B6396; BL 1171. e. 20. (2.)

81 CARPENTER, Agricola
**Pseuchographia Anthropomagica: or, A Magicall Description of the Soul: wherein is set forth the nature, genesis and exodus of it.* Pp. 25.
London, for John Browne
1652
8° [4°, Wing; 12°, BL]
Wing C613; BL E. 1369. (5.)

CARPENTER, Richard
Worke (see ASHMOLE, *TCB*)

82 CASE, John
Wards of the Key to Helmont proved Unfit for the Lock; or, the Principles of Mr. W. Bacon examined and refuted, and the honour and value of true chymistry asserted. Pp. 24.
London, for the Author, sold by John Smith
1682
4°
Wing C821; BL 1034. h. 36.

83 CASTLE, George
Chymical Galenist: a treatise wherein the practise of the ancients is reconcil'd to the new discoveries in the theory of physick . . . In which are some reflections upon a book [by M. N., i.e., Marchamont Nedham], *intituled Medela Medicinae.* Pp. 196.
London, Sarah Griffin for Henry Twyford and Timothy Twyford
1667
8°
Wing C1233; BL 1038. f. 25.

84 CATTAN, Christophe de
The Geomancie of Maister Christopher Cattan . . . Whereunto is annexed the wheele of Pythagoras. Translated out of French into our English tongue (by Francis Sparry).
London
1591: John Wolfe to be sold by Ed. White, 4° STC 4864; BL 1609/784
1608: E. A[llde] for Ed. White, 4° STC 4865; BL 719. e. 44.

85 CHALONER, Thomas
A Shorte Discourse of the most rare and excellent vertue of Nitre: wherein is declared the . . . cures by the same effected. ff. 22.
London, printed for Gerard Dewes
1584
4°
STC 4940; BL 778. e. 56.

CHAMBERLAYNE, John
**The Manner of Making Coffee, Tea, and Chocolate* (see DUFOUR, Sylvestre)

CHAMBERLEN, Peter
(see *Publick Bathes purged, or a Reply to Dr. Chamberlens Vindication of Publick Artificial Bathes . . .*)

86 CHAMBERLEN, Peter
A Vindication of Publick artificiall Baths and Bath-stoves, from the Objections and Scandalls of some of the Colledge of Physitians that do not or will not know the benefit of them to the Publick. Pp. 6.
London
1648
4°
Wing C1909; BL 551. b. 9. (2*.)

87 CHAPPUZEAU, Samuel
The History of Jewels and of the principle riches of the East and West. Taken from the relation of divers of the most famous travellers of our age, translated from the French of C[HAPPUZEAU].
London, T. N. for Hobart Kemp
1671
8°
Wing C1959; BL 954. b. 39.

88 CHARAS, Moise
**Royal Pharmacopoea, Galenical and chymical, according to the practice of the most eminent and learned physitians of France.*
London, for John Starkey and Moses Pitt
1678
fol.
Wing C2040; BL 547. l. 11.

CHARLETON, Walter (see HELMONT, Jean Baptiste van, *Deliramenta Catarrhi*)

CHARNOCK, Thomas
Breviary of Natural Philosophy, and *Aenigma's* (see ASHMOLE, *TCB*)

CHAUCER, Geoffrey
Chanon's Yeoman's Tale (see ASHMOLE, *TCB*)

89 CHILDREY, Joshua
Britannia Baconica, or Natural rarities of England, Scotland, and Wales. According as they are to be found in every shire, and according to the Precepts of my Lord Bacon; Methodically digested, with Observations, whereby divers secrets in Nature are discovered. Pp. 184.
London, printed for the author; sold by H. E.
1661
8°
Wing C3871; BL 291. b. 38.

Chymical Aphorisms (see HELMONT, Franciscus Mercurius)

Chymical Dictionary (see HEYDON, John, *The English Physician's Guide*, and SENDIVOGIUS, Michael, *A New Light of Alchymie*)

90 CLARKE, William
Natural History of Nitre: or, a Philosophical discourse of the nature, generation, place, and artificial extraction of nitre, with its vertues and uses. Pp. 93.
London, E. Oakes for Nat. Brook
1670
8°
Wing C4564; BL 987. a. 12.

91 C[LERKE], **T**[imothy], Dr.
Some Papers writ in the Year 1664 in Answer to a Letter [signed: T. M., i.e., Christopher Merret], *concerning the Practice of Physick in England.* By Dr. C. T. [i.e., Christopher Terne]. Pp. 51.
London, for James Allestry
1670
4°
Wing T760; BL 1171. 1. 45. (3.)

[Wing lists under Terne, Christopher.]

92 CLEVER, William
Flower of Phisicke. Wherein is perfectlie comprehended a true introduction and method for mans assured health . . . with three bookes of philosophie for the due temperature of mans life. Pp. 122.
London, Roger Ward
1540 [1590, STC, BL]
4°
STC 5412; BL 774. e. 5. (1.)

93 CLOWES, William
**A Short and Profitable Treatise touching the Cure of the Disease called Morbus Gallicus, etc.* [Contains G. Baker's *The Nature and Propertye of Quicksilver.*]
London, J. Daye
1575 [1579, STC, BL]
8°
STC 5447; BL 1174. a. 23.

94 COELSON [COLSON], **Lancelot**
Philosophia maturata: an Exact Piece of Philosophy, containing the Practick and Operative part thereof in gaining the Philosophers Stone; with the wayes how to make the Mineral Stone, and the Calcination of Mettals. With the work of St. Dunstan concerning the Philosophers Stone, and the Experiments of Rumelius, and the preparation of Angelo Sala. Pp. 142.
London, G. Sawbridge
1668
12°
Wing C4883; BL 1036. a. 49.

Collectanea Chymica (see COOPER, William)

95 COLLINS, John
Salt and Fishery, a discourse thereof. Pp. 164.
London, by A. Godbid and J. Playford; sold by Robert Horne
1682
4°
Wing C5380; BL 981. b. 5.

96 COLLY, Anthony

Natures Champion, sounding a challenge to her stoutest assailants; or, a more ample explanation of the virtue and use of my pilulae aureae purgantes, whose operation is Hemetick, Purgative, Diaphoretick, Dieuretick, Anodyne, Narcotick, &c. Pp. 40.
London, printed for the author
1670
4°
Wing C5395B; BL T. 991. (1.)

97 COLMENERO DE LEDESMA, Antonio
**Chocolate: or, an Indian drinke. By the wise and moderate use whereof, health is preserved, sicknesse diverted, and cured . . . Written originally in Spanish . . . and faithfully rendred* [sic] *in the English, by Capt. James Wadsworth.* Pp. 40. [Another edition of *A Curious Treatise.*]
London, J. G. for John Dakins
1652
8°
Wing C5400; BL E. 1671. (1.)

98 COLMENERO DE LEDESMA, Antonio
**A Curious Treatise of the Nature and Quality of Chocolate . . . put into English by Don Diego de Vades-forte* [i. e. James Wadsworth]. Pp. 21.
London, J. Okes
1640
4°
STC 5570; BL T. 339. (8.)

99 COLONNA, Francesco
**Hypnerotomachia. The Strife of Loue in a Dreame,between Polyphilius and Polio, being accounted a Treatise of the Philosophers Stone.* [With Woodcuts.] Ff. 100.
London, J. Charlewood, for S. Waterston
1590 [1592, STC, BL]
4°
STC 5577; BL Huth. 39.

[A translation of the first book of *Hypnerotomachia Poliphili* of Francesco Colonna. The Translator's dedication signed: R. D. (Sir Robert Dallington?)]

100 COMBACHIUS, Ludwig
Fundamenta Chymica, or a sure Guide into the high and rare Mysteries of Alchimy.

London, Wm. Godbid for William Barlow
1657 [1658, Wing]
8°
Wing C5436A

[Cooper states, "... which is a new Title only put to Ludwig Combachius his Salt, Lumen, and Spiritus Mundi Philosophici." (see HESTEAU, Clovis, *Sal, lumen, et spiritus mundi philosophici*)]

101 COMENIUS, Johann Amos
Naturall Philosophie reformed by Divine Light, or a Synopsis of Physics; being a view of the World in general . . . With a briefe appendix touching the diseases of the body, mind, and soul; with their generall remedies.
London, by Robert & Wm. Leybourne for Thomas Pierrepont
1651
8°
Wing C5522; BL E. 1335.

A Conference concerning the Phylosophers Stone
(see HARTLIB, *CMCA*)

102 COOPER, William
A Catalogue of Chymicall Books. In Three parts. In the First and Second Parts are contained such Chymical Books as have been written Originally, or Translated into English: With a large Account of their Titles, several Editions and Volumes. Likewise In the Third Part is contained a Collection of such things published in the Philosophical Transactions of the Royal Society (for Ten Years together) as pertain to Chymistry, or the Study of Nature by Art in the Animal, Vegetal, and Mineral Kingdoms. (For an earlier edition, see C[OOPER], W[illiam], *Philosophical Epitaph.*)
London, T. R. & N. T. for William Cooper
1675
8°
Wing C6061; BL G. 19086.

103 C[OOPER], **W**[illiam]
**Collectanea Chymica: a collection of Ten Several Treatises in Chymistry, concerning the Liquor Alkahest, the Mercury of Philosophers, and other Curiosities worthy the Perusal. Written by Eir. Philaletha, Anonymous, Joh. Bapt. Van-Helmont, Dr. Fr. Antonie, Bernhard Earl of Tre-*

visan, Sir Geo. Ripley, Rog. Bacon, Geo. Starkey, Sir Hugh Platt, and the Tomb of Semiramis, the epistle to the reader signed: W. C. B., [i.e., William Cooper, Bookseller].
London, Printed for William Cooper, at the Pelican in Little Britain
1684
8°
Wing C5103; BL 1033. d. 15.

["Notwithstanding the Latin titles the tracts are all in English, except the first, which is in both languages. Each tract has a title-page," Ferguson, I, 169.]

Contents:

1. **Eirenaeus PHILALETHA**, *Arcanum, or Secret of the Immortall Liquor Alkahest*, p. 5.
2. **ANONYMOUS**, *Practice of Lights*, p. 27.
3. **Joh. Bapt. Van HELMONT**, *Praecipiolum*, p. 47.
4. **Fr. ANTONIE**, *Aurum Potabile*, p. 73.
5. **BERNARD of Trevisan**, *Treatise*, p. 83.
6. **George RIPLEY**, *Bosome-book*, p. 101.
7. **Roger BACON**, *Speculum Alchymiae*, p. 125.
8. **George STARKEY**, *Oil of Sulphur-vive*, p. 137.
9. **Hugh PLATT**, *Remedies against Famine*, p. 153.
10. **H. V. D.**, *Tumba Semiramidis*.

104 C[OOPER], **W**[illiam]
**Continuation or Appendix to the Second Part of the Catalogue of Chymical Books, printed with The Philosophical Epitaph, being either Corrections, Rectifications, or Additions to the Editions of the old Books; and a Continuation of such New Books as have been Printed since the Year 1675. to this present Year 1688.*
1688
8°
BL 270. b. 16.

[On this *Continuation*, see the Introduction to this edition.]

105 C[OOPER], **W**[illiam] (?)
The Philosophical Epitaph of W. C. Esquire, for a Memento Mori on his Tomb-stone.
London, T. R. & N. T. for William Cooper
1673
8°

Wing C6062; BL 1033. g. 22. (1.)

Contents:
1. *Three Hieroglyphical Scutcheons displaying Minerva's and Hermes Birds, and Apollo's Bird of Paradise in Philosophical Mottos and Sentences with their Explication, and a discovery of the Liquor Alchahest, of Salt of Tartar volatized, and other Elixirs, with their differences.*

2. *A Briefe of the Golden Calfe, discovering the rarest Miracle in Nature, of a transmutation of Lead into Gold made by Dr. Jo. Fred. Helvetius with Figures.*

3. *Jo. Rod. Glauber his Golden Ass well managed, and Midas restored to Reason, a new Chymical light for comfort of the oppressed, demonstrating Gold to be easily extracted in all places out of Sand, Stones, Gravel, or Flints, and the true matter of the Philosophers Stone.*

4. *Jehior, the day Dawning, or the Morning Light of Wisdom, containing the three Principles or Originals of all things whatsoever, discovering the great and many Mysteries in God, Nature, and the Elements.*

All published by W. C., Esq. With a *Catalogue of Chymical Books* (collected by Will. Cooper, Bookseller)

106 C[OOPER], **W**[illiam] (?)
Secrets of alchimy.
London
8°

[This entry appears only in the First Part of Cooper's *Catalogue*; on its importance, see the Introduction to this edition.]

107 COTTA, John
Cotta Contra Antonium: or, an Ant-Antony, or an Ant-Apology, manifesting Doctor Antony his Apologie for Aurum potabile, in true and equall ballance of right Reason, to be false and counterfait. Pp. 108.
Oxford, John Lichfield & James Short, for Henry Cripps.
1623
4°
STC 5832; BL 1142. h. 30.

108 COTTA, John
A Short Discoverie of the unobserved dangers of severall sorts of ignorant and unconsiderate Practisers of Physicke in England: Profitable not onely for the deceived multitude, and easie for their meane capaci-

ties. Pp. 135.
London, William Jones & Richard Boyle
1612
4°
STC 5833; BL 551. a. 2. (1.)

109 COX, Fr.
**Treatise of the making and use of diverse Oyls, &c. and stilled Waters.*
London
1575
8°

[Additional information unavailable.]

110 COXE, Thomas or D[aniel]
Discourse, wherein the interest of the patient in reference to physick and physicians is soberly debated. Many abuses of the apothecaries in the preparing their medicines are detected, and their unfitness for practice discovered. Pp. 333. (see *Brevis Responsio*)
London, for Richard Chiswell
1669
8°
Wing C6727; BL 1038. e. 26.

CREMONENSIS, Gerhard
Of Astronomical Geomancy (see AGRIPPA, *Fourth Book*)

111 CROLLIUS, Oswald
Admonitory Preface of Oswald Crollie, Physitian: to the most illustrious Prince Christian Anhaltin. [A Translation of Crollius, *Basilica Chymica*, by H. Pinnell, under the title, *Philosophy reformed and improved in four profound tractates: The I. discovering the great and deep mysteries of Nature: by that learned chymist & physitian Osw: Crollius. The other III. discovering the wonderfull mysteries of the creation, by Paracelsus: being his philosophy to the Athenians*]. 4 pt.
London, by M. S. for Lodowick Lloyd
1657
8°
Wing C7023; BL E. 1589. (1.)

112 CROLLIUS, Oswald
Bazilica chymica, & Praxis chymiatricæ. Or, Royal and practical chymistry, in three treatises . . . Being a translation of Oswald Crollius his

Royal Chymistry, augmented and inlarged by John Hartman. To which is added his Treatise of Signatures of Internal Things . . . As also The Practice of Chymistry of John Hartman M. D. augmented and inlarged by his son. All faithfully Englished by a lover of chymistry. 3 pt.
London, for John Starkey, and Thomas Passinger
1670
fol.
Wing C7022; BL C. 113. h. 8.

CULPEPER, Nicholas, translator (see Partlitz, *New Method*)

113 CULPEPER, Nicholas
**Culpeper's School of Physick. Or the Experimental practice of the whole art. Wherein are contained all inward diseases from the head to the foot, with their proper and effectuall cures, such diet set down as ought to be observed in sickness or in health . . . The narrative of the authors life is prefixed, with his nativity calculated, etc.* Pp. 361.
London, for N. Brooke
1659
8° [4°, Wing]
Wing C7544; BL E. 1739.

114 CULPEPER, Nicholas
Mr. Culpepper's Treatise of Aurum Potabile; Being a description of the three-fold world. Viz. elimentary, celestiall, intellectuall, containing the knowledge necessary to the study of hermetick philosophy.
London, for G. Eversden
1656
8° [12°, Wing]
Wing C7549; BL 1032. a. 9.

115 CYRANUS, King of Persia
**The Magick of Kirani King of Persia, and of Harpocration; containing the magical and medicinal vertues of stones, herbs, fishes, beasts and birds . . . Now published and translated into English from a copy found in a private hand.* Pp. 156.
London, for W. Cooper [T. C. II, 202]
1685
8°
Wing M249; BL 1141. a. 15.

D., B. Gent.

A Rare and excellent discourse of minerals (see PLATT, Sir Hugh, *Jewell House of Art and Nature*, 1653 edition)

D., H. V., *Tumba Semiramidis* (see COOPER, *Collectanea Chymica*)

DASTIN, John
Dastin's Dream (see ASHMOLE, *TCB*)

116 DEANE, Edmund
Spadacrene Anglica. Or, the English spaw-fountaine. Being a brief treatise of the acide, or tart fountaine discovered, by Mr. Slingsbey in the Forest of Knaresborow in the West-Riding of Yorkshire. Pp. 32.
London, John Grismand, sold by R. Foster in Yorke
1626
4°
STC 6441; BL 1171. d. 45.

117 DEE, Arthur
Fasciculus chemicus; or, Chymical collections. Expressing the ingress, progress, and egress, of the secret Hermetick science, out of the choisest & Most famous authors . . . Whereunto is added, The Arcanum or Grand secret of Hermetick philosophy [by Jean d'Espagnet]. *Made English by James Hasolle* [Elias Ashmole]. Pp. 268.
London, J. Flesher for Richard Mynne
1650
8°
Wing D810; BL 234. a. 18.

DEE, John
Testament (see ASHMOLE, *TCB*)

118 DERHAM, Samuel
**Hydrologia Philosophica; or, an Account of Ilmington Waters in Warwickshire, with directions for drinking of the same. Together with some experimental observations touching the original of compound bodies.* Pp. 162.
Oxford, by Leon Litchfield for John Howell
1685
8°
Wing D1098; BL 234. h. 54.

Dictionary (see *Physical Dictionary*)

119 DIGBY, Sir Kenelm
Choice and Experimented Receipts in Physick and Chirurgery; as also cordial and distilled waters and spirits, perfumes, and other curiosities.
London, printed for the Author
1668
8°
Wing D1423; BL 1651/14.

120 DIGBY, Sir Kenelm
**A Choice Collection of Rare Secrets and Experiments in Philosophy . . . Collected and experimented by . . . Sir Kenelm Digby . . . now . . . published by Mr. Hartman his Steward and Operator.*
London, to be sold by W. Cooper, Henry Farthorne, and John Kersey
1682 and reissued
8°
Wing D1425A, D1426, D1426A; BL C. 112. aa. 14.

121 DIGBY, Sir Kenelm
Discourse concerning the Vegetation of Plants. Spoken by Sir Kenelme Digby, at Gresham College, on the 23. of January, 1660. At a meeting of the Society for promoting Philosophical Knowledge by Experiments. Pp. 100.
London, J. G. for John Dakins
1661
12° [24°, BL]
Wing D1432; BL 1651/708.

122 DIGBY, Sir Kenelm
A Late Discourse made in a solemne assembly of nobles and learned men at Montpellier in France . . . Touching the cure of wounds by the powder of sympathy; with instructions how to make the said powder; whereby many other secrets of nature are unfolded . . . Rendered faithfully out of French into English by R. White, Gent. The second edition corrected and augmented, with the addition of an index. Pp. 152.
London, R. Lowndes; T. Davies
*1658: [first and second editions], 8° [12°, Wing, BL] Wing D1435/ D1436; BL 1038. a. 23.
1660: [third edition], 8° [12°, Wing, BL] Wing D1437; BL 1038. a. 25.

123 *A Discourse of Magical Gold.*

[Cooper refers to this title only in the First Part of the *Catalogue* and provides no additional information; title not located in STC, Wing, or BL.]

A Discourse of subterraneal Treasure
(see L[AWRENCE], T[homas], *Mercurius Centralis*)

Distillation of all Manner of Spices, Seeds, Roots and Gums
(see *True and perfect order . . .*)

124 **Distiller of London.*
London, by Robert Paske, for the Company of Distillers
(see FRENCH, John, *Art of Distillation*)
1668
fol.
Wing D1692; BL 7953. i. 1.

125 DUCHESNE, Joseph [QUERCETANUS]
A Breefe Aunswere of J. Quercetanus . . . to the exposition of J. Aubertus Vindonis concerning the original and causes of Mettalles. Set foorth against Chimists. Another . . . treatise . . . concerning the Spagericall preparations, and use of minerall, animall, and vegitable medicines. Whereunto is added divers rare Secretes. . . . By J. Hester, etc. 2 pts.
London, sold by J. Hester
1591
8°
STC 7275; BL 1033. b. 3.

126 DUCHESNE, Joseph [QUERCETANUS]
Practise of Chymicall and Hermeticall Physicke, for the preservation of health, . . . written in Latin by J. Quersitanus . . . and translated into English by T. Timme, 2 pt.
London, T. Creede
1605
4°
STC 7276; BL 1039. g. 4. (1.)

127 DUCHESNE, Joseph [QUERCETANUS]
The Sclopetarie of J. Quercitanus . . . or his booke containing the cure of Wounds received by shot of gunne or such like engines of warre. Englished by J. Hester. (see also FIORAVANTI, *Three Exact Pieces* and PARACELSUS, *A hundred and fourteene Experiments*)
London, R. Ward for J. Sheldrake

1590
4° [8°, BL]
STC 7277; BL 783. d. 8. (1.)

DUCHESNE, Joseph [QUERCETANUS]
Spagyrick Antidotary (see DUCHESNE, *Sclopetarie*, and FIORAVANTI, *Three Exact Pieces*)

128 DU CLOS, Samuel Cotreau, sieur
[or COTTEREAU DU CLOS, Samuel]
**Observations on the Mineral Waters of France, made in the Royal Academy of the Sciences . . . Now made English.* Pp. 125.
London, for Henry Faithorne, and John Kersey
1684
8° [12°, BL]
Wing D2432; BL 233. a. 42.

129 DUDLEY, Dud
D. D.'s Metallum Martis: or, iron made with pit-coale, sea-coale, etc., and with the same fuell to melt and fine imperfect mettals, and refine perfect mettals.
London, by T. M. for the Author
1665
8°
Wing D2438; BL 446. a. 25. (1.)

130 DUFOUR, Sylvestre, [SPON, Jacob]
**The Manner of Making Coffee, Tea, and Chocolate. As it is used in most parts of Europe, Asia, Africa, and America. With their vertues. Newly done out of French and Spanish* [by J. Chamberlayne]. [The tracts on tea and coffee translated from the French of Jacob Spon, and that on chocolate from the Spanish of Antonio COLMENERO DE LEDESMA.]
3 pts. Pp. 116.
London, for William Crook
1685
12°
Wing D2454; BL 449. a. 7.

DUNSTAN, St.
The Philosophers Stone (see COELSON, *Philosophia Maturata*)

131 DU PLESSIS [**or DUPLEIX**], **Scipion**
The Resolver; or Curiosities of nature, useful and pleasant for all.
[Translated from the French.]
London, N. & J. Okes
1635
12°
STC 7362; BL 8706. a. 12.

132 ELSHOTZ [**or ELSHOLTZ**], **Johann Sigismund**
**The Curious Distillatory; or the art of distilling coloured liquors, spirits, oyls, . . . put into English by F. S*[herley].
London, by J. D. for Robert Boulter
1677
8°
Wing E638; BL 1038. f. 27. 2.

133 ERCKER, Lazarus
**Fleta Minor. The laws of art and nature in knowing, judging, assaying . . . metals. In two parts. The first contains assays by L. Ercker in V. books . . . translated into English. The second contains essays on metallick words alphabetically composed as a dictionary. By Sir J. Pettus. Illustrated with 44 sculptures.* 2 pts.
London, for the author, by Thomas Dawks
1683
fol.
Wing P1906; BL L. 35/69.

ESPAGNE, Jean d' [ESPAGNET]
Arcanum, or the grand secret of Hermetick Philosophy; wherein the secrets of Nature and Art, concerning the Matter and Manner of making the Philosophers Composition are orderly and methodically manifested
(see DEE, Arthur, *Fasciculus Chemicus*)

134 ESPAGNE, Jean d' [ESPAGNET]
Enchyridion Physicae Restitutae, or Summary of Physicks' Recovery, wherein the true Harmony of Nature is explained, and many errors of the Ancient Philosophers by Canons and Demonstrations evidenced and evinced, [translated by Dr. Johann Everard.]
London, by W. Bentley, to be sold by W. Sheares and Robert Tutchein
1651
12°
Wing E3276A

135 *An Essay for the Regulation of the Practice of Physick for the composure of all differences between Physitians and Apothecaries, &c.*
London, for Tho. Taylor
1673
4°
Wing E3283; T. C. I, 134.

**Essayes of natural experiments made in the Accademia del Cimento*
(see FLORENCE, Accademia del cimento)

136 EVANS, John
Universal Medicine; or, the vertues of the Antimoniall Cup. Collected out of the experiments and observations of the most famous, learned, and best approved philosophers and physicians that have written of that subject.
London
1634: J. Haviland, 8° STC 10587; BL 7383. a. 21.
*1642: by Richard Hodgkinsonne, 4° Wing E3452; BL 778. e. 41. (3.)
*1651: by Richard Hodgkinsonne, 4° Wing E3452A

[*Cooper reports, "Dr. Primrose answered this book in his *Popular Errors*"; see PRIMROSE, James.]

137 EVELYN, John
**A philosophical discourse of Earth, relating to the culture and improvement of it for vegetation, and the propagation of plants, etc.* Pp. 182.
London, for John Martyn
1676
8°
Wing E3507; BL 967. c. 5.

EVONYMOUS (see GESNER, Conrad)

An Explanation of Ripleys Epistle (see HARTLIB, *CMCA*)

138 FABER, Albertus Otto
Paradoxon de morbo Gallico. Libr. II. or, a Paradox concerning the shameful disease, forwarning to all against deceitful cures . . . Translated out of the High Dutch by J. Kauffman.
London
1662
12°
Wing F67; BL 1174. a. 39.

139 FABER, Albertus Otto
Some kindling sparks in matters of physick to satisfie some physicians who are of opinion, that spirits (which they call hot things) do burn and inflame the body.
London
1668
4°
Wing F70; BL 1034. h. 24. (1.)

FABER, Peter John
**112 Chymicall Arcanums*
(see SALMON, William, *Polygraphicae*)

FELGENHAUER, Paul
Jehior, Aurora sapientia, the Day-break of wisdom
(see COOPER, William, *Philosophical Epitaph*)

140 FENTON, Edward
Certaine secrete wonders of Nature, containing a description of sundry strange things, seeming monstrous in our eyes and judgement, because we are not privie to the reasons of them; with Figures, [Translated by E. F. from the *Histoires Prodigieuses* of P. Boaistuau]. Pp. 148.
London, H. Bynnemann
1569
4°
STC 10787; BL C. 31. d. 8.

141 FIORAVANTI, Leonardo
A Compendium of the rationall secretes of the . . . moste excellent Doctour of Phisicke . . . L. Phioravante . . . devided into three Bookes. [Translated and edited by J. Hester.]
London
*1582: J. Kyngston for G. Pen and J. H., 8° STC 10879; BL 1507/1846.
1652 [1653]: [as *An Exact Collection of the choicest . . . experiments*], by G. Dawson, to be sold by John Saywell, 4° Wing F951B
*1659: for William Shears, 4° Wing F952

142 FIORAVANTI, Leonardo
**A Joyfull Jewell. Contayning . . . orders, preservatives . . . for the Plague . . . written in the Italian tung by . . . L. Fioravantie . . . and now . . . translated . . . by T. H*[ill]. [Edited by J. Hester.]
London, for W. Wright

1579
4°
STC 10880; BL C. 31. e. 5.

143 FIORAVANTI, Leonardo
**A short discours . . . uppon chirurgerie . . . whereunto is added a number of notable secretes . . . translated out of Italyan into English by J. Hester.*
London
1580: T. East, 4° STC 10881; BL 549. b. 24. (3.)
1626: E. Allde, 4° STC 10882; BL 7480. b. 7.

144 FIORAVANTI, Leonardo
Three exact pieces of L. Phioravant, . . . Viz. his rationall secrets, and chirurgery, reviewed and revived . . . Whereunto is annexed Paracelsus his one hundred and fourteen experiments; with certain excellent works of B. G. à Portu Aquitano; Also Isaac Hollandus his secrets concerning his Vegetall and animall work; with Quercitanus his spagyrick antidotary [Sclopetarie] *for Gun-shot.* 4 pt.
London, by G. Dawson to be sold by Wm. Nealand
1652
4°
Wing F953; BL E. 642.

145 FLAMEL, Nicolas
His exposition of the hieroglyphicall figures which he caused to bee painted upon an Arch in St. Innocent's Church-yard in Paris . . . Together with the secret booke of Artephius and the epistle of John Pontanus: concerning both the theoricke and the practicke of the Philosophers Stone. Faithfully, . . . and religiously done into English out of the french and latine copies by Eirenaeus Orandus.
London, T. S. for T. Walkley
1624
12°
STC 11027; BL 1032. a. 19.

FLAMEL, Nicolas
Summary of Philosophy (see HOUPREGHT, *Aurifontina Chymica*)

146 FLETCHER, Richard
**A character of a true Physician; or a true chymist compared with a goose quill pedant, etc.*

London, for the author
1678 [1676, Wing, BL]
8°
Wing F1357; BL 778. c. 39.

147 FLETCHER, Richard
**Starkey revived; or, collections out of nature's explication and Helmont's vindication. pt. I.*
London, for the author
1676
8°
Wing F1359; BL 780. b. 3. (4.)

148 FLORENCE, ACCADEMIA DEL CIMENTO
**Essayes of natural experiments made in the Accademie del Cimento . . . Englished by R. Waller.*
London, for Benjamin Alsop
1684
4°
Wing A161; BL 434. g. 6.

149 FLUDD, Robert
**Dr. Fludd's Answer unto M. Foster. Or the squesing of Parsons [sic] Fosters Sponge, ordained by him for the wiping away of the Weapon-Salve.*
London, printed for N. Butter
1631
4°
STC 11120; BL 1038. 1. 3. (2.)

150 FLUDD, Robert
Mosaicall Philosophy, grounded upon the essentiall truth or eternal sapience.
London, for Humphrey Moseley
1659
fol. [4°, BL]
Wing F1391; BL C. 66. f. 1.

151 FOSTER, William
**Hoplocrisma-Spongus; or, a Sponge to wipe away the Weapon-Salve. A Treatise, wherein is proved, that the Cure late-taken up amongst us, by applying the Salve to the Weapon, is Magicall and unlawfull.* Pp. 56.

London, by Thomas Coates, for John Grove
1631
4°
STC 11203; BL 1038. l. 3. (1.)

152 *The free Exchanger or the Mint for the uncurrant Coyne, according to the true Value of grains wanting in light Gold, with the several worth of all pieces of Gold, at this present.*
London
1627
8°

[Additional information unavailable.]

153 FRENCH, John
Art of Distillation; or, a treatise of the choicest spagyricall preparations . . . together with the description of the chiefest furnaces and vessels used by ancient and moderne chymists, and the Anatomy of Gold and Silver, with their preparations, curiosities, and virtues; with two Books of Sublimation and Calcination.
London
1632: [no record in STC, BL, or Wellcome]
*1639: [no record in STC, BL, or Wellcome]
1651: by Richard Cotes to be sold by Thomas Williams, 4° Wing F2169; BL E. 619. (5.)
*1653, 52: [*The Art of distillation . . . the second edition. To which is added, The London-Distiller: exactly and truly shewing the way . . . to draw all sorts of spirits and strong-waters.* 2 pts.] by E. Cotes, for Thomas Williams, 4° Wing F2170; BL 1651/732.
*1664: [*The Art of Distillation . . . To which is added in this third impression Calcination and sublimation . . . As also, the London-distiller, exactly and truly shewing the way . . . to draw all sorts of spirits and strong waters.*] E. Cotes for T. Williams, 4° Wing F2171; BL 1651/979.
1667: [*The Art of distillation: . . . To which is added in this fourth impression, Sublimation and calcination, in two books; As also, The London Distiller.*] E. Cotes for T. Williams, 4° Wing F2172; BL 1035. e. 3.
*1668: (see *The Distiller of London*)

154 FRENCH, John
York-shire Spaw; or, a treatise of four famous medicinal wells, (viz) The Spaw, or Vitrioline; the Sting, or Sulphur; the Dropping, or Putrifying;

and S. Magnus Wells in Yorkshire, their Cause, Virtue, and use.
London, for Nath. Brook
1654
8° [4°, Wing]
Wing F2176; BL G. 3100.

155 FULKE, William
**A Goodly Gallerye with a most pleasaunt Prospect, into the garden of naturall Contemplation, to behold the Naturall causes of all kynde of Meteors, as well fyery and ayery, as watry and earthly.*
London, W. Griffith
1563
8°
STC 11435; BL b. 54. a. 25.

156 FULLER, Thomas
**History of the Worthies of England.*
London, by J. G. [,] W. L. and W. G.
1662
fol.
Wing F2440; BL N. L. 18. c.

[Cooper's specific reference is to the "History of the silver Works and Mines in Cardiganshire in Wales, pag. 3."]

157 GAFFAREL, Jacques
Unheard of curiosities concerning the Talismanical sculpture of the Persians, the Horoscope of the Patriarcks, and the reading of the Stars.
[Englished by E. Chilmead.]
London, G. D. for H. Moseley
1650
8°
Wing G105; BL E. 1216. (1.)

GALEN
Booke of Elements (see JONES, John, *Briefe . . . discourse*)

GALID, the son of Jazich
Secrets of Alchimy (see BACON, Roger, *Mirror of Alchimy*)

158 GARENCIÈRES, Theophilus de
**The admirable virtues and wonderful effects . . . of tincture of coral in physic, etc.*

London, by W. R. for Samuel Sprint
1676
8°
Wing G253; BL 778. c. 33.

GEBER
The Works of Geber (see JĀBIR IBN HAIYĀN)

159 GEDDE, Walter
**A booke of sundry draughtes, principaly serving for Glasiers; and not impertinent for Plasterers, and Gardiners: . . . whereunto is annexed the manner how to anniel in glas: And also the true forme of the fornace, and the secretes thereof.* 2 pt.
London, W. Dight
1615–16
4°
STC 11695; BL C. 21. b.[2] (1–2.)

160 GELLIBRAND, Henry
Discourse Mathematical of the Variation of the Magneticall Needle together with its admirable diminution lately discovered.
London, W. Jones
1635
4°
STC 11712; BL C. 112. b. 8.

161 GESNER, Conrad [EVONYMOUS]
A new booke of destillatyon of waters, called the Treasure of Evonymous . . . whereunto are joyned the forms of sondry apt furnaces . . . Translated by P. Morwyng.
London, J. Day
1565
4°
STC 11801; BL 462. c. 11.

162 GESNER, Conrad [EVONYMOUS]
Newe Jewell of Health, wherein is contained the most excellent Secretes of Phisicke and Philosophie; and of all the Distillations of Waters, Oyles, Balmes, Quintescences; with the Extraction of Artificial Salts, the use and Preparation of Antimony, and Potable Gold, with the Vessels and Furnaces, and other Instruments thereunto belonging; Being the Second part of the Treasury of Evonymus [Trans. by George Baker].

London, H. Denham
1576
4°
STC 11798; BL C. 113. b. 1.

163 GESNER, Conrad [EVONYMOUS]
Practise of the New and Old Physicke, being his Jewel of health, or Distillations . . . Newly corrected and published in English by G. Baker.
London, P. Short
1599
4°
STC 11799; BL 46. m. 2.

164 GLASER, Christophe
**Compleat Chymist, or a New Treatise of Chymistry . . . from the fourth edition revised and augmented by the Author. Now faithfully englished by a Fellow of the Royal Society.* Pp. 285.
London, for John Starkey
1677
8°
Wing G843; BL 1142. a. 36.

165 GLAUBER, John Rudolph
Description of new Philosophical Furnaces, or, a new art of distilling, . . . Whereunto is added a description of the Tincture of Gold, or the true Aurum Potabile, with the first part of the Mineral Work.
London, by Ric. Cotes for T. Williams
1651
4°
Wing G846; BL E. 649. (3.)

GLAUBER, John Rudolph
Golden Ass well managed (see COOPER, *Philosophical Epitaph*)

166 GLAUBER, John Rudolph
**The Works of the Highly Experienced and Famous chymist, John Rudolph Glauber: containing, Great Variety of Choise Secrets in Medicine and Alchemy, In the working of the Metallick Mines, and the Separation of Metals: . . . Translated into English by Christopher Packe.*
London, by Thomas Milbourn, for the author
1689
fol.

Wing G845: BL 545. 1. 11.

167 GODDARD, Jonathan
Discourse setting forth the unhappy condition of the practice of physick in London.
London, by John Martyn and James Allestry
1669 [1670, Wing, BL]
4°
Wing G914; BL 1172. h. 1. (5.)

168 GODFREY, Robert
Various injuries and abuses in chymical and galenical physick committed both by Physicians and Apothecaries, detected, &c.
London, by John Darby, for Richard Jones
1674
8°
Wing G927; BL 1038. a. 17.

GOWER, John
Upon the Philosophers Stone (see ASHMOLE, *TCB*)

GRATAROLE, William
Of the Philosophers Stone (see P., H., *Five Treatises*)

169 GREW, Nehemiah
Anatomy of Vegetables begun. With a general account of vegetation.
London, for Spencer Hickman
1671 [1672, Wing, BL]
12° [8°, Wing, BL]
Wing G1946; BL 987. e. 20.

170 GREW, Nehemiah
**The Comparative Anatomy of Trunks; together with an account of their vegetation, grounded therupon; in two parts.*
London, by J. M. for Walter Kettilby
1675
8°
Wing G1947; BL 987. f. 3.

171 GREW, Nehemiah
**A Discourse made before the Royal Society, December 10. 1674. Concerning the Nature, Causes, and Power of Mixture.*

London, for John Martyn
1675
12°
Wing G1948; BL 1651/1134.

172 GREW, Nehemiah
Idea of a phytological history propounded. Together with a continuation of the anatomy of vegetables, particularly prosecuted upon roots, with seven Figures.
London, by J. M. for Richard Chiswell
1673
8°
Wing G1951; BL 449. c. 2.

173 GUIDOTT, Thomas
**A discourse of Bathe, and the hot waters there. Also some enquiries into the nature of the water of St. Vincent's Rock near Bristol; and that of Castle Cary. To which is added a century of observations . . . With an account of the lives and character of the Physicians of Bathe.* Pp. 200.
London, for H. Brome
1676
8°
Wing G2192; BL 7470. bb. 26.

174 GUIDOTT, Thomas
A discourse of natural bathes . . . with . . . an Appendix concerning Bathe, wherein the Antiquity both of the Bathes and City is more fully discoursed; with a brief account of the Nature and Virtues of the Hot-waters there. (see JORDEN, Edward)
London, to be sold by Tho. Salmon in Bathe
1669
8°
Wing G2189; BL 1171. f. 10.

175 H., J.
A Treatise of the great antidote of Van Helmont, Paracelsus, and Crollius; by them called the Elixir proprietatis: . . . and an examination of author's [sic] *concerning its true preparation; as also, a ready way to volatile salt of tartar, etc.* Written by J. H. [John Hodgkins.]
London, for the Author
1667: 4° Wing H82; BL 7510. b. 27.
*1668: 4° Wing H82A

*1674 [1671, Wing]: 4° Wing H82B

176 [HALE, Sir Matthew]
Difficiles Nugae: or, observations touching the Torricellian experiment, and the various solutions of the same, touching the weight and elasticity of air.
London, by W. Godbid for William Shrowsbury [sic]
1674
8°
Wing H238; BL 1136. d. 4.

177 [HALE, Sir Matthew]
An Essay touching the Gravitation or Non-gravitation of Fluid Bodies, and the reason thereof.
London, by W. Godbid for William Shrewsbury
1673
8°
Wing H244; BL 538. a. 19.

178 HARRIS, Walter
**Pharmacologia anti-empirica, Or a rational discourse of remedies, both Chymical and Galenical; . . . with some remarks on the causes and cure of the Gout, the universal use of the cortex, or Jesuits powder, etc.*
London, for Richard Chiswell
1683
8°
Wing H885; BL 547. c. 9.

179 HARTLIB, Samuel
Chymical, Medicinal and Chyrurgical Addresses, made to Samuel Hartlib, Esquire, In nine Treatises. [*CMCA*]
London, By G. Dawson for Giles Calvert
1655
8° [12°, BL]
Wing H978; BL 1036. a. 37.

Contents:
1. *Whether the Urim and Thummim were given in the Mount, or perfected by Art.*
2. Sir Geo. Ripley's *Epistle to King Edward the Fourth unfoulded.*
3. Gab. Plattes' *Caveat for Alchymists.*
4. *A Conference concerning the Phylosophers Stone.*

5. *An Invitation to a free and generous Communication of Secrets and Receits in Physick.*
6. *Whether or no each several Disease hath a particular Remedy.*
7. *A new and easie Method of Chirurgery for the curing of all fresh Wounds, or other hurts.*
8. *A Discourse about the Essence or Existence of Metals* by Gerh. Maylnes.
9. *The new Postillions pretended prophetical Prognostication of what shall happen to Physitians, Chyrurgians, Apothecaries, Alchymists and Miners.*

HARTMAN, John
Royal Chimistry (see CROLLIUS, *Bazilica Chymica*)

180 HARVEY, Gideon
Accomplisht Physician, the honest Apothecary and the skilful Chirurgian, detecting their necessary connexion and dependence on each other. Withall a Discovery of the frauds of the Quack Emperick, the prescribing Surgeon, and the Practising Apothecary . . . Whereunto is added . . . a lash for Lex Talionis. Pp. 95.
London
1670
4°
Wing M1835; BL 551. a. 42.

181 HARVEY, Richard
**An astrologicall discourse upon the . . . conjunction of Saturne & Jupiter which shall happen the 28 day of April 1583. With a . . . declaration of the effects which the late eclipse of the Sunne 1582 is yet heereafter to worke.*
London, H. Bynnemann
1583
8°
STC 12910; BL C. 112. a. 15.

[HASOLLE, James] i.e., **ASHMOLE, Elias**
(see DEE, Arthur, *Fasciculus Chymicus*)

182 HELMONT, Franciscus Mercurius van
**Cabbalistical Dialogue concerning the Worlds Creation, in answer to . . . a learned Doctor* [i.e., Henry More], *. . . To which is subjoined a . . . paraphrastical exposition of Genesis I. . . . by the author of the di-*

alogue, etc.
London, for Benj: Clark
1682
4°
Wing H1390; BL 701. g. 47.

183 HELMONT, Franciscus Mercurius van
**One Hundred Fifty Three Chymical Aphorisms . . . by Eremita Suburbanus* [i.e., F. M. van Helmont], . . . *To which are added some other Philosophick Rules or Canons*, [by Ber. G. Penotus a Portu Aquitano], *pertaining to his Hermetick Science . . . Made English by C. Packe.*
London, for the author, to be sold by W. Cooper; and D. Newman
1688
12°
Wing H1392; BL 1035. a. 33. (2.)

[BL attributes this title to J. B. van Helmont.]

184 HELMONT, Franciscus Mercurius van
**The Paradoxal Discourses of F. M. van Helmont concerning the Macrocosm and Microcosm of the greater and lesser World, and their union, set down in writing by J. B. and now published.* 2 pt.
London, by J. C. and Freeman Collins, for Robert Kettlewel
1685
8°
Wing H1393; BL 784. b. 20.

185 HELMONT, Jean Baptiste van
Deliramenta Catarrhi; or, the incongruities, impossibilities, and absurdities couched under the vulgar opinion of defluctions; with Dr. Charleton's paraphrase upon the same. Pp. 75.
London, E. G. for W. Lee
1650
4°
Wing H1398; BL E. 601. (6.)

186 HELMONT, Jean Baptiste van
Oriatrike, or Physick refined, the common errors therein refuted, and the whole art reformed and rectified: being a new rise and progress of philosophy and medicine [with the preface of the original editor F. M. van Helmont] . . . *rendered into English* . . . by J. C[handler].
London
*1662: for Lodowick Lloyd, fol. Wing H1400; BL 549. l. 26.

1664: [reissued as van Helmont's *Workes*] for Lodowick Lloyd, fol. Wing H1397; BL 549. l. 27.

HELMONT, Jean Baptiste van
**Praecipiolum* (see C[OOPER], W[illiam], *Collectanea Chymica*)

187 HELMONT, Jean Baptiste van
Ternary of Paradoxes of the Magnetic Cure of wounds, the nativity of tartar in wine, and the image of God in Man, translated by Dr. Walter Charleton.
London, J. Flesher for W. Lee
1650
4°
Wing H1401; BL E. 582. (1.)

HELVETIUS, Johann Friedrich
Briefe of the Golden Calfe (see COOPER, *Philosophical Epitaph*)

188 HELVETIUS, Johann Friedrich
Golden Calf, which the World adores, and desires. In which is handled the most rare . . . wonder of Nature, in transmuting Metals.
London, for John Starkey
1670
8° [12°, Wing, BL]
Wing H1410; BL 8907. a. 24.

189 HENSHAW, Nathaniel
Aero-chalinos: or, a Register for the Air . . . for the better preservation of health and cure of diseases, after a new method.
1664: Dublin, for Samuel Dancer, 8° Wing H1481; BL C. 112. a. 7. (5.)
*1677: [second edition], London, for Benj. Tooke, 12° Wing H1482; BL 717. a. 51. (1.)

190 HERMANNI, Philippus
**An Excellent Treatise teaching howe to cure the French Pockes; with all other diseases arising and growing thereof, and in a manner all other sicknesses. Drawn out of the bookes of . . . Theophrastus Paracelsus. Compiled by . . . P. H. . . . and now put into English by J. Hester.*
London, J. Charlwood
1590
4°
STC 13215; BL 1174. b. 4.

191 HERMES, Trismegistus
**The Divine Pymander of Hermes Mercurius Trismegistus, in XVII Books. Translated formerly out of the Arabick into Greek, and thence into Latin, and Dutch, and now out of the original into English; by that learned Divine Doctor Everard.* [With a preface, signed J. F.] Pp. 215.
London, R. White, for T. Brewster and G. Moule
1650
8°
Wing H1565; BL E. 1344. (2.)

192 HERMES, Trismegistus
**Hermes Mercurius Trismegistus, his Divine Pymander. . . . Together with his second book, called Asclepius . . . with a commentary. Translated formerly out of the Arabick into Greek, and thence into Latine and Dutch, and now out of the original into English; by . . . Dr. Everard.* [The preface signed J. F.] 72 pt.
London, by J. S. for T. Brewster
1657
12°
Wing H1567; BL 8460. aa. 15.

HERMES, Trismegistus
Iatromathematica; Physical Mathematics and Mathematical Physics (see WILLIAMS, Ralph, *Physical Rarities*, second edition)

HERMES, Trismegistus
Smaragdine Table of Alchimy (see BACON, Roger, *Mirror of Alchimy* and P., H., *Five Treatises*)

193 HESTEAU, Clovis, Sieur de Nuysement
**Sal, lumen, et spiritus mundi philosophici: or, the Dawning of the day, discovered by the beams of light: shewing the true salt and secret of the philosophers, the first and universal spirit of the world. Written originally in French [by Baron J. de Nuisement], afterwards turned into Latin . . . by Lodovicus Combachius . . . and now transplanted into Albyons garden, by R. T.* [Robert Turner]. Pp. 220.
London, by J. C. for Martha Harrison
1657
8°
Wing N1469; BL C. 145. f. 20.

HESTER, John, translator

A Breefe Aunswere of J. Quercetanus (see DUCHESNE, Joseph)

HESTER, John, translator
An excellent Treatise how to cure the French Pockes (see HERMANNI, Philippus)

HESTER, John, translator
A hundred and foureteene Experiments and Cures (see PARACELSUS, *A hundred and foureteene Experiments*, and FIORAVANTI, *Three exact pieces*)

194 HESTER, John
**The Pearle of Practise; or, Practiser's Pearle, for Physicke and Chirurgerie. Found out by J. H[ester] . . . Since his death . . . garnished and brought into some methode by a welwiller of his* [J. Fourestier].
London, R. Field
1594
4°
STC 13253; BL 543. b. 23. (1.)

HESTER, John, editor
Preservatives and pretious Practices against the Plague (see FIORAVANTI, Leonardo, *A Joyfull Jewell*)

HESTER, John, translator
The Sclopeterie of J. Quercitanus (see DUCHESNE, Joseph)

HESTER, John
Secretes (see DUCHESNE, Joseph, *A Breefe Aunswere*)

HESTER, John, translator
The . . . Secretes of Phisicke and Philosophie, with the Order of the Distillation of Oyls, etc. (see PARACELSUS, *Key of Philosophie*, 1633)

HESTER, John, translator
A short discours . . . uppon chirurgerie (see FIORAVANTI, Leonardo)

195 HEYDON, John
El Havareuna; or, the English Physitians Tutor in the Astrobolismes of Mettals Rosie Crucian, Miraculous Saphiric Medicines of the Sun and the Moon; the Astrolasmes of Saturn, Jupiter, Mars, Venus, Mercury, &c. 2 pt.

London, for William Gilbertson
1665, 64
8°
Wing H1669; BL C. 66. b. 1.

196 HEYDON, John
English Physitians Guide, or the Holy Guide: leading the way to the Wonder of the World (a Compleat Physician) teaching the knowledge of all things, past, present, and to come, viz. of pleasure, long life, health, Youth, Blessedness, Wisdome and Virtue . . . with Rosie Crucian Medicines, etc. . . . where unto is added, a Bar to stop Thomas Street from his . . . mad . . . clambering up to Astronomy, . . . "with a small Chymical Dictionary," etc. 6 pt.
London, T. M. for Samuel Ferris
1662
8°
Wing H1666; BL 1034. a. 8. (1, 2.)

197 HEYDON, John
Hampaaneah Hammeguleh: or, the Rosie Crucian Crown, the second and third Books, set with seven Angels, seven Planets, seven Genii, twelve Signs, twelve Ideas, sixteen Figures, and their occult powers upon the seven Metals, and their miraculous Virtues in Medicines; with the perfect and full discovery of the Pantarua and Elixirs of Metals prepared to cure Diseases; whereunto is added Elhauareuna preforio, Regio Lucis & Pfonthon.
London, for the Author
1665 [1664, Wing]
8°
Wing H1667A

[*Hampaaneah Hammeguleh* is included in Book 2 (1665) and Book 3 (1664) of the British Library copy (1607/4633) of Heydon's *Wise Mans Crown.*]

198 HEYDON, John
Harmony of the World: being a discourse wherein the phaenomena of nature are consonantly solved and adapted to inferiour intellects.
London, for H. Brome
1662
8°
Wing H1668; BL 8631. a. 37.

199 HEYDON, John
**The Idea of the Law charactered from Moses to King Charles. Whereunto is added the Idea of Government and Tyranny.* 3 pt.
London, for the author
1660
8°
Wing H1671; BL E. 1916. (2.)

200 HEYDON, John
**A New Method of Rosie Crucian Physick, wherein is shewed the cause, and therewith their experienced medicines for the cure of all diseases, etc.*
London, for Thomas Lock
1658
4°
Wing H1672; BL E. 946. (4.)

HEYDON, John
**Ocia Imperialia: being select exercises of philosophy, war, government, &c.* (see HEYDON, John, *Theomagia*)

201 HEYDON, John
**The Rosie Crucian infallible Axiomata, or Generall Rules to know all Things past, present, and to come, etc.*
London
1660
8°
Wing H1674; BL 719. a. 36.

[Cooper adds, "The same is in the second book of his Holy Guide. 1662. 8°."]

202 HEYDON, John
Theomagia, or the Temple of Wisdome. In three parts, spiritual, celestial, and elemental, containing the occult powers of the Angels of Astromancy . . . The Mysterious vertues of the characters of the stars . . . The Knowledge of the Rosie Crucian Physick, and the miraculous secrets in Nature. [Includes *Ocia Imperialia: being select exercises of philosophy, war, government,* &c.] 3 pt.
London, by T. M. for Henry Brome and Tho. Rooks
1664–3
8°
Wing H1676; BL 8632. b. 49.

203 HEYDON, John
Wise Mans Crown: or, the glory of the Rosie-Cross, shewing the wonderful power of nature; with the full discovery of the true Coelum Terrae, or first matter of Metals, and their preparations into Elixirs that cure all diseases; with the Regio Lucis, or holy household of Rosie Crucian Philosophers.
London, for the author by Samuel Speed
1664
8°
Wing H1677; BL 1607/4633.

204 HIGHMORE, Nathaniel
History of Generation; examining the several opinions of divers Authors, especially that of Sir K. Digby, in his Discourse of bodies . . . To which is joyned, A discourse of the Cure of Wounds by sympathy . . . especially by that powder known chiefly by the name of Sir Gilbert Talbot's Powder.
London, by R. N. for John Martin
1651
8°
Wing H1969; BL E. 1369. (3.)

HIPPOCRATES
A Discourse of Humane Nature (see *Key to unknowne Knowledge*)

HIPPOCRATES
Hippocrates chymicus (see TACHENIUS, Otto)

HOLLAND, Isaiah [Isaac]
Secrets concerning his Vegetall and animall works (see FIORAVANTI, *Three Exact Pieces* and PARACELSUS, *A hundred and foureteene Experiments*)

HOLLAND, Isaiah [Isaac]
Work of Saturn (see VALENTINE, Basil, *Of Natural and Supernatural Things*)

205 HOOKE, Robert
Micrographia, or some physiological descriptions of minute bodies made by magnifying glasses; with observations and inquiries thereupon.
London, J. Martyn and J. Allestry
1665
fol.

Wing H2620; BL 435. e. 19.

HORTULANUS
Commentary upon the Smaragdine Table of Hermes (see BACON, Roger, *Mirror of Alchimy*)

206 HOTHAM, Charles
**An Introduction to the Teutonick Philosophie. Being a determination concerning the Original of the Soul. . . . Englished by D. F.*
London, by T. M. & A. C. for Nath. Brooks
1650
24° [12° , Wing, BL]
Wing H2896; BL 527. a. 30.

207 HOUGHTON, Thomas
**The Compleat Miner; or, a Collection of the Laws, Liberties, ancient Customs . . . and Privileges of the several Mines and Miners in the Counties of Derby, Gloucester, and Somerset; together with the Art of dialling, etc.* 3 pt.
London, for William Cooper
1688–87
12°
Wing H2926; BL 517. a. 24. (1–3.)

[Pt. 1 also published as *Rara Avis in Terris*, 1681; pts. 2 and 3 published separately in 1687 as *The Laws . . . of the Miners in the Forrest of Dean* and *The Ancient Laws . . . of the Miners in the King's Forrest of Mendipp.*]

208 HOUGHTON, Thomas
**Rara Avis in Terris; or, the Compleat Miner, . . . containing the Liberties, Laws and Customs of the Lead-mines . . . in Derbyshire, . . . the Art of dialling and levelling grooves . . . With an explanation of the Miners terms of Art.*
London, printed for William Cooper
1681
12°
Wing H2934; BL 445. a. 25. (1.)

209 HOUPREGHT, John Frederick
**Aurifontina Chymica, or, a Collection of Fourteen small Treatises Concerning the First Matter of Philosophers, for the discovery of their (hitherto so much concealed) Mercury.*

London, for William Cooper
1680
24° [12°, BL]
Wing H2941; BL 1036. a. 29. [lost]; Wellcome 39689.

Contents:
1. *Hydropyrographum Hermeticum, of the Metallick Water-Fire, translated by J. F. HOUPREGHT*, p. 1.
2. *The Privy Seal of Secrets, plainly discovering the first Matter of the Philosophers*, p. 41.
3. *A strange letter concerning an Adept, his curious Learning and vaster Treasure, from Frederick, Duke of Holsatia and Sleswick*, p. 53.
4. **Sir George RIPLEY's** *Treatise of Mercury, and the Philosophers stone*, p. 69.
5. *Colours to be observed in the operation of the Great Work of Philosophers*, p. 93.
6. *A plain and true Description of the Treasure of Treasures, or the Golden Medicine*, p. 97.
7. *A Treatise of the Philosophers Stone, the Blessed Manna, &c., with its wondrous Virtues and Use, both for the curing of the Body of Man, and the making of the precious Stones*, p. 107.
8. **Nic. FLAMMELL**, *his Summary of Philosophy, or Short Treatise of the Philosophers Stone*, p. 145.
9. **Raym. Lullie's** [*LULL*] *Clavicula, Apertorium, or little Key explaining all the rest of his Works*, p. 163.
10. *Secrets disclos'd of the Philosophers Stone*, p. 180.
11. *A Philosophical Riddle of Gold, in Verse*, p. 185.
12. **Bern. TREVISAN's** *Epistle to Thomas of Bononia, concerning the secret workings of Nature in the product of things*, p. 187.
13. **Bern. TREVISAN**, *his short Epistle Dedicatory before his Works, to the same man*, p. 269.
14. *A brief Preparation of the Philosophers Stone, for the conclusion of this Book*, p. 271.

210 HUGHES, William
The American physitian; or, a treatise of the roots, plants, trees, shrubs, fruit, herbs, &c., growing in the English Plantations in America, &c., whereunto is added a discourse of the Cacao-nut-tree, and the use of its fruit, with all the ways of making Chocolate.
London, by J. C. for William Crook
1672
12°
Wing H3332; BL 449. a. 24.

Hydropyrographum Hermeticum (see HOUPREGHT, *Aurifontina Chemica*)

211 IRVINE, Christopher, translator
Medicina Magnetica: or, the rare and wonderful Art of Curing by Sympathy; laid open by Aphorismes; proved in conclusions; and digested into an easy method [translated from the Latin of William Maxwell by Christopher Irvine].
Edinburgh, C. Higgins [London, BL]
1656
8°
Wing I1053; BL E. 1578. (1.)

212 JĀBIR IBN HAIYĀN
The Works [*or rather the Summa Perfectionis, Liber investigationis magisterii, De inventione veritatis and Liber Fornacum*] *of Geber, the most famous Arabian Prince and Philosopher . . . Englished by R. Russel.*
London
1678: for N. E. by Thomas James; to be sold by Robert Clavel, 8° Wing J54; BL 236. i. 19.
1686: for William Cooper, 8° Wing J55; BL 1507/329

Jehior. The Day dawning or the Morning light of Wisdom, containing the three Principals or Originals of all things (see COOPER, William, *Philosophical Epitaph*)

213 JONES, John
A briefe . . . discourse of the naturall beginning of all growing and living things: heate, generation; . . . use and abuse of phisicke, preservation, etc. . . . Hereunto is annexed a . . . work, entituled, Galen's booke of Elements, . . . in the ende wherof is adjoyned two other bookes ("The Bathes of Bathes Ayde," "The Benefit of the auncient bathes of Buckstones") containing the nature of all the Bathes in England, etc.
London, W. Jones
1574–72: 4° STC 14725; BL C. 31. b. 38. (1–4.)

JONES, John
Translation of Galen's Booke of Elements in the Epitome or the originall of all things* (see **JONES, *John, A briefe . . . discourse*)

214 JONSTON, John
History of the Constancy of Nature. Wherein, by comparing the latter

age with the former, it is maintained that the world doth not decay universally, . . . nor man in his age, stature, strength, or faculties of his minde, as relating to all arts and science. By John Jonston of Poland.
London, for John Streater
1657
8°
Wing J1016; BL E. 1653. (1.)

215 JONSTON, John
History of the wonderful things of Nature, of Heaven, the Elements, Meteors, Minerals, Plants, Birds, Beasts, . . . Rendered into English by a Person of Quality.
London, by John Streater
1657
fol.
Wing J1017; BL 461. i. 14.

216 JORDEN, Edward
Discourse of naturall Bathes, and Minerall Waters . . . especially of our Bathes at Bathe in Sommersetshire.
London, T. Harper
1631: 4° STC 14791; BL 234. h. 29.
1632: [enlarged second edition], 4° STC 14792; BL 1171. h. 18. (1.)

217 JORDEN, Edward
**Discourse of naturall Bathes* [third edition], *. . . enlarged; with some particulars of the author's life. To which is added an appendix concerning Bathe . . . By T. Guidott, M. B.* 2 pt.
London, T. Harper, sold by M. Sparke
1633
4°
STC 14793 [Wing J1074 is "third edition", 1669, 8°, sold by Thomas Salmon in Bathe. This is BL 1171. f. 10.]

218 JORDEN, Edward
Discourse of naturall Bathes [fourth edition] *with Eugenius Philanders Quere concerning drinking Bath-water at Bathe, resolved; with an Appendix concerning Bathe, and the Antiquity both of the Bathes and City; with the Natures and Virtues of Hot-waters; by Thomas Guidott.* Pp. 26.
London, for George Sawbridge, and Thomas Salmon, in Bathe
1673
8°

Wing J1075; BL G. 19189.

219 JOSSELYN, John
New England's rarities discovered: in birds, beasts, fishes, serpents and Plants of that country. Pp. 114.
London, G. Widdowes
1672
8°
Wing J1093; BL 435. a. 5.

KELLEY, Edward
Work of the Philosophers Stone (see ASHMOLE, *TCB*)

220 KENDALL, George
An Appendix to the Unlearned Alchimist [of Richard Mathews], *wherein is contained the true Receipt* [for the preparation and making of] *Matthew's Pill,* (see MATHEWS, Richard, *Unlearned Alchymist*).
London, for Joseph Leigh to be sold by the author
1663 [1664, Wing, BL]
8°
Wing K283; BL 1033. d. 2. (2.)

221 **Key to unknowne Knowledge; or, A Shop of five Windows ... consisting of five . . . treatises: 1. The Judgment of urines. 2. Judiciall rules of Physike. 3. Questions of Oyles. 4. Opinions for curing of Harquebush-Shot. 5. A Discourse of humane Nature* [by Hippocrates].
London, A. Islip for E. White
1599
4°
STC 14946; BL 774. f. 1.

222 KEYSER, Johann von Breda
Batavia's Praecipitatum, or East-India Mercury unexpected praecipitatum, or the death of all deluded Philosophers and Seekers of the Lapis Philosophorum; with a Concept or Compositio Saturni to make and know the true Mercurium, the Argentum Vivum; with the knowledge and concent of [*P. P.*] *sent from the Island Java to Amsterdam.*
London
1673
4°

[Not found in Wing or BL; BL 1033. c. 65. is a German translation, Hamburg?, 1681, 12°, of the work that Cooper lists.]

223 KHRYPFFS, Nicolaus
**The Idiot in four books; the first and second of wisdome, the third of the minde, the fourth of statick experiments, etc. By the famous and learned C. Cusanus* (Translated from the Latin).
London, for Wm. Leake
1650
12°
Wing K394; BL 1248. a. 34.

224 L[awrence], **T**[homas]
Mercurius centralis, or, a discourse of subterraneal Treasure, occasioned by some late discoveries thereof, in the County of Norfolk, and sent in a Letter to Dr. Browne.
London, for J. Collins
1668
12° [8°, Wing]
Wing L689B

225 LAMBI, Giovanni Battista
Revelation of the secret spirit. Declaring the most concealed secret of Alchymie. Written first in Latine by an unknowne author, but explained in Italian by John Baptista Lambye. Translated by R. N. E. gent.
London, J. Haviland for H. Skelton
1623
16° [8°, STC]
STC 15184; BL C. 124. aaa. 20.

226 **The Laws and Statutes of the Stannary or Tin Mines of the County of Devon . . . containing the Charters granted to the Tinners . . . by King Edward I. II. III. IV. and V. by King Henry VIII, Edward VI and Q. Elizabeth, etc.*
London, V. Simmes
1600
fol.
STC 6798; Folger

227 LE FEVRE, Nicolas
A compendious body of Chymistry; teaching the whole practice thereof by the most exact preparation of animals, vegetables, and minerals, preserving their essential vertues . . . Rendered into English by P. D. C., Esq.
London, for Thomas Davis and Theo. Sadler
1662

4°
Wing L924A

228 LE FEVRE, Nicolas
Compleat body of chymistry: . . . Rendred into English, by P. D. C., Esq. . . . with additions. 2 pt.
London
1664: by Tho. Ratcliffe for Octavian Pulleyn, junior, 4° Wing L925; BL 1651/1384.
1670: for O. Pulleyn, junior, and are to be sold by John Wright, 4° Wing L926; BL 1651/1415.

229 LE FEVRE, Nicolas
Discourse upon Sir W. Rawleigh's Great Cordial . . . Rendered into English by P. Belon.
London, by J. F. for Octavian Pulleyn, junior
1664
8°
Wing L928; BL 1034. f. 35.

230 LEMERY, Nicolas
**An appendix to A course of chymistry, being additional Remarks to the former Operations, together with the Process of the volatile Salt of Tartar.* Pp. 140.
London, printed for Walter Kettilby
1680
8°
Wing L1037A; BL 1651/1036. (2.)

231 LEMERY, Nicolas
**A course of chymistry. Containing the easiest manner of performing those operations that are in use in physick . . . Translated by Walter Harris.* Pp. 323.
London
1677: printed for Walter Kettilby, 8° Wing L1038; BL 1651/1036 (1.)
*1686: [second edition enlarged] by R. N. for Walter Kettilby, 8° Wing L1039; BL 8905. aa. 21.

232 LEMERY, Nicolas
**Modern Curiosities of Art and Nature. Extracted out of the cabinets of the most eminent personages of the French Court. Together with the choicest secrets in mechanics . . . Made English from the original*

French.
London, printed for Matthew Gilliflower, and James Partridge
1685
12°
Wing L1041; BL 1489. cc. 84.

233 LEMNIUS, Levinus
Secret Miracles of Nature. In four books . . . Whereunto is added one book containing . . . rules how man shall become excellent in all conditions, etc.
London, by Jo. Streater, to be sold by Humphrey Moseley, John Sweeting, John Clark, and George Sawbridge
1658
fol.
Wing L1044; BL 441. f. 2.

Letter . . . Concerning an Adept (see HOUPREGHT, *Aurifontina Chymica*)

234 *A Letter sent by a learned Physitian to his friend, wherein are detected the manifold Errors used hitherto of the Apothecaries, in preparing their Compositions, as Syrups, Condites, Conserves, Pills, Potions, Electuaries, Lozenges, &c; with a far better manner to prepare and correct the same.*
London
1586
8°

[Entry in Maunsell is identical; no further information found.]

235 LEVENS, Peter
A right profitable Booke for all diseases, called . . . the Path way to health, for Distilling of divers Waters, and making of Oyls &c.
London
1587: Edward Allde for Edward White, 4° STC 15533; BL C. 31. d. 24.
1664: for J. W. to bee sold by Charles Tyus, 12° Wing L1819A; BL E. 1039. f. 26.

236 *Little Book of Secrets for liquifying and using of Gold and Silver.*
London
8°

[No additional information found, but possibly an edition of *Booke of*

Secrets.]

The London Distiller (see FRENCH, John, Art of Distillation)

237 LOVELL, Robert
Παμβοτανολογια, sive enchiridion botanicum. . . . Compleat herball, or the sum of Galenical and Chymical Authors, touching Trees, Shrubs, Plants, Fruits, Flowers, &c.
Oxford
*1659: by William Hall for Ric. Davis, 8° [12°, BL] Wing L3243; BL E. 1858, 59.
1665: [second edition enlarged], by W. H. for Ric. Davis, 12° Wing L3244; BL 988. a. 9.

238 LOVELL, Robert
Πανζωορυκτολογια. Sive panzoologico-mineralogia . . . Compleat history of animals and minerals; Being the sum of Ancient and Modern Galenical & Chymical Authors concerning Beasts, Birds, Fishes, Serpents, Insects, and Man; and of Earths Metals, Semi-metals, Salts, Sulphurs, and Stones, both Natural and Artificial; With their place, matter, names, kinds, temperature, virtues, use, choise, &c. 2 pt.
Oxford, by Hen: Hall for Joseph Godwin
1661
8°
Wing L3246; BL E. 1810, 11.

LULL, Raymond
Clavicula (see HOUPREGHT, *Aurifontina Chemica*)

LULL, Raymond
Philosophical and Chymical Experiments (see PARACELSUS, *Chymical Transmutation*)

239 LUPTON, Thomas
**A Thousand Notable things of sundry sorts, being a collection of many curious Secrets, after the manner of Alexis, where of some are wonderful, some strange, some pleasant.*
London
1627: E. All-de for N. Fosbrooke, 4° [8°, BL] STC 16960; BL 236. k. 29.
1670: by Ed. Crowch for John Wright, 8° Wing L3499
1686: for M. Wotton and G. Conyers, 8° Wing L3501; BL 546. c. 35.

LYDGATE, John
Secreta Secretorum, or Letter of Alexander the Great to Aristotle (see ASHMOLE, *TCB*)

240 M., T. [MERRET, Christopher]
Letter to a Doctor [subscribed, T. M., i.e., C. Merret] *in London concerning the present state of physic, and the Regulation of the practice of it in this kingdom, for the Advancement of the knowledge of Nature in general.*
London, for Jo. Martyn and Ja. Allestry
1665
4°
Wing M81C; BL 1171. l. 45. (2.)

241 MACKAILE, Matthew
**The Diversitie of Salts and Spirits maintain'd. . . . By way of Animadversions upon Dr. D. C*[oxe] *his three papers . . . in the 9. vol. of the P*[*hilosophical*] *T*[*ransactions*] *. . . As also Scurvie Alchymie dicovered.*
Aberdeen, by John Forbes
1683
8° [12°, BL]
Wing M144; BL 1142. a. 35.

242 MACKAILE, Matthew
**Moffet-well; or, a topographical-spagyricall description of the mineral wells at Moffet . . . Translated and . . . enlarged by the author. As also, the Oyly Well, . . . at St. Catherine's Chappel in the Paroch of Libberton. To these is subjoined a character of Mr Culpepper and his Writings.*
Pp. 196.
Edinburgh, for Robert Brown
1664
8°
Wing M148; BL 1171. e. 43.

MADEIRA ARRAIS, Edward
(see BACON, Roger, *The Cure of Old-Age*)

**Magic of Kirani* (see CYRANUS, King of Persia)

The Magistery of W. B. (see ASHMOLE, *TCB*)

243 *Magnetical Philosophy*

8°

[Cooper includes this reference only in the First Part of the *Catalogue* and provides no additional information; title not located in STC, Wing, or BL.]

244 MAIER, Michael [MAJERUS]
Lusus serius: or, serious passe-time. A philosophical discourse concerning the superiority of creatures under man. [Translated by John Hall.]
London, for Humphrey Moseley and Tho. Heath
1654
12°
Wing M286; BL 970. a. 1.

245 MAIER, Michael [MAJERUS]
Themis Aurea. The Laws of the Fraternitie of the Rosie Cross. Written in Latin . . . now in English. Pp. 136
London, for N. Brooke
1656
12° [8°, Wing, BL]
Wing M287; BL G. 16137.

MALYNES, Gerhard
Philosophical Discourse about the Essence or Existence of Metals (see HARTLIB, *CMCA*)

246 MANLOVE, Edward
Treatise of Darbyshire Miners. A Poem. The Liberties and Customes of the Lead-mines within the wapentake of Wirksworth in the County of Derby . . . Composed in meeter. Pp. 9.
London
1653: 4° Wing M453; BL 517. b. 19.
1657: for Thomas Johnson, 4° Wing M453A

247 MAPLET, John
A greene Forest, or a naturall Historie, Wherein may be seene . . . the most Sufferaigne Vertues in all the whole kinde of Stones and Mettals,. . . . Plants, Herbs, Trees, . . . and Shrubs, of Beasts, Fowls, Fishes, creeping Worms, and Serpents.
London, H. Denham
1567
8°
STC 17296; BL 957. a. 5.

248 MARSHALL, William
**Answers upon several heads in Philosophy.* (The first section . . . conteining the first six answers upon several heads in philosophy. The second section . . . conteining queries relating to the angle of contact.)
London, by T. L. for Nathaniel Brooke
1670
8°
Wing M809A; BL 536. a. 21.

249 MARSHALL, William
**Philosophy Delineated, conteining a resolution of divers knotty questions upon suddry [sic] philosophical notions, etc.* Pp. 237.
[Second edition of *Answers upon several heads in Philosophy.*]
London, for Obadiah Blagrave
1678
8°
Wing M809B; BL 1393. a. 36.

250 MASCALL, Leonard, translator
**A Profitable boke declaring dyvers approoved remedies, to take out spottes and staines, in Silks, Velvets, . . . with divers Colours how to die . . . also to dress Leather, etc. Englished by L. M*[ascall].
London, T. Purfoote
1596: 4° STC 17592
1605: 4° STC 17593; BL C. 31. c. 20.

251 MATHEWS, Richard
Unlearned Alchymist his Antidote, or a more full and ample Explanation of the use, virtue and benefit of his Diaphoretick and Diuretick Pill; together with a precious Pearle in the midst of a Dunghill, or the preparation of the Oyl of Amber, Oyl and Balsom of the Pine-tree; with a universal Purge of Antimony and Scammony, of Cream or Christals of Tartar, with their Correction, Preparation and Use, with the preparation of Crocus Martis, Sugar of Saturn, Essence of Venus, Salt of Mars, Sympathetical Powder, &c.
London
1660: for Joseph Leigh, to be sold by Giles Calvert and Livewell Chapman, 8° Wing M1290
*1662: for Joseph Leigh and by Giles Calvert, 8° [12°, BL] Wing M1290B; BL 780. a. 16.
*1663: for Joseph Leigh, 8° Wing M1291; BL 1651/1186 or 1033. d. 2. (1).
1664: (see KENDALL, George, *An Appendix*)

***MAXWELL, William**
Medicina Magnetica (see IRVINE, Christopher)

252 MAYNWARING, Everard
Medicus Absolutus, . . . Compleat Physician, qualified and dignified. The rise and progress of physick, Historically, Chronologically, and Philosophically illustrated, Physitians of different Sects & Judgements charactered and distinguished, &c.
London, for the booksellers
1668
8°
Wing M1497; BL 1038. ff. 51.

253 MAYNWARING, Everard
Nova medendi ratio, a short and easie method of curing. Exemplified by a ternary of radical medicines, universal in their respective classes.
London, by J. Dover, for Henry Broome
1666
4°
Wing M1508; BL 1172. i. 5. (2.)

254 MAYNWARING, Edward
Praxis Medicorum antiqua et nova: the ancient and modern practice of physick examined, stated and compared; the preparation of Medicines according to the primitive Custome of Princely Physitians asserted and proved, the grand duty of every Physitian, &c.
London, by J. M. to be sold by T. Archer
1671
4°
Wing M1512; BL 551. a. 9. (4.)

255 MAYNWARING, Edward
Vita sana et longa. [i.e., *Pharmacopaeian Physitians Repository*]. *The preservation of health, and prolongation of Life, proposed and proved in the due observance of remarkable precautions, and duly practicable Rules, relating to body and mind, compendiously abstracted from the Institutions and Laws of Nature.* 2 pt.
London, by J. D.
1669
8°
Wing M1519; BL 1039. e. 24.

256 MERRET, Christopher
Reflections on a Libel [by Jer. Richardson] *intituled, A Plea for the Apothecaries.*
London, for Richard Chiswell
1671
4°
Wing C165; BL 551. a. 7.

[The *Continuation* notes, "by a Mistake D. Dan Foot's name was (in the second part of this catalogue) put in, instead of Dr. Merrett, the known author thereof."]

257 MERRET, Christopher
Self-conviction, or an Enumeration of the Absurdities, Railings against the College, and Physicians in general; . . . and also an answer to the rest of Lex Talionis. (see STUBBE, Henry)
London, for James Allestry
1670
4°
Wing M1841; BL 551. a. 43.

258 MERRET, Christopher
Short Reply to the Postscript [to *Campanella Revived*], *etc., of H. S*[tubbs] *shewing his many falsities in matters of fact; the impertinencies of his promised answers to some physicians that have written against the Apothecaries, his conspiracy with the Apothecaries to defame them.* (see STUBBE, Henry)
London, by T. R., for J. Allestry
1670
4°
Wing M1842; BL 550. b. 11.

259 MERRET, Christopher
Short view of the frauds and abuses committed by Apothecaries . . . and of the only remedy thereof by Physicians making their own medicines.
London
1669: [first edition], for James Allestry, 4° Wing M1843; BL 551. a. 12. (1.)
1670: [second edition], for James Allestry, 4° Wing M1844; BL 551. a. 45.

260 **Dr. Meverell's Answers to the Lord Bacon's Questions, concerning the Compounding, Incorporating, Separation, Variation, and Resti-*

tution of Metals and Minerals in the Lord Bacon's Physiological Remains [in *Baconiana. Or Certain Genuine Remains of Sr. Francis Bacon*].
London, by J. D. for Richard Chiswell
1679
8°
Wing B269; BL 535. b. 1.

261 MOELLENBROCK, Valentin Andreas
**Cochlearia curiosa: Or the Curiosities of Scurvygrass, being an exact scrutiny and careful Description of the Nature and medicinal Virtue, . . . Written in Latine by Dr. V. A. Molimbrochius . . . Englished by Thomas Sherley.*
London, S. and B. Griffin for Wm. Cademan
1676
8°
Wing M2381; BL 957. i. 22.

262 MONARDES, Nicolas
Joyfull Newes out of the newe founde worlde, wherein is declared the rare and singuler vertues of diverse and sundrie Hearbes, Trees, Oyles, Plantes, and Stones, . . . And the portrature of the saied hearbes . . . Englished by J. Frampton. Pp. 109.
London
1577: W. Norton, 4° STC 18005A; BL 968. e. 19/1.
1596: [another edition], E. Allde, by the assigne of B. Norton, 4° STC 18007; BL 236. 1. 42.

263 MORE, Henry
**Conjectura Cabbalistica: or, a conjectural essay of interpreting the minde of Moses according to a three-fold Cabbala, viz. literal, philosophical, mystical, or, divinely moral.* Pp. 251.
London, J. Flesher, to be sold by William Morden, in Cambridge
1653
8°
Wing M2647; BL E. 1462. (2.)

264 MORE, Henry [PHILALETHES, Alazonomastix]
Observations upon Anthropososophia Theomagica and Anima Magica abscondita [by Eugenius Philalethes, i.e., T. Vaughan]. pp. 94.
Parrhesia [London], O. Pullen
1650
8°
Wing M2667; BL 526. f. 14. (3.)

[In his citation of More's attack upon Thomas Vaughan in the *Observations*, Cooper refers to a second attack included "with Dr. More's *Atheismus Triumphatus*. 1656 8°." Doubtless intending to refer to More's *Enthusiasmus Triumphatus* [London: J. Flesher, 1656, 8° Wing M2655], Cooper has probably confused Tommaso Campanella's *Atheismus Triumphatus* (1631) and More's own *An Antidote Against Atheism* (1653).]

265 MORE, Henry
**Ψυχωδια Platonica; or, a platonicall song of the soul, consisting of foure severall poems;. . . . Hereto is added a paraphrasticall interpretation of the answer of Apollo, consulted by Amelius, about Plotinus soul departed this life.*
Cambridge, by Roger Daniel
1642: 8° Wing M2674; BL E. 1108.
1647: 8° Wing M2647A

266 MORE, Henry [PHILALETHES, Alazonomastix]
The Second Lash of Alazonomastix; conteining a Solid and Serious Reply to a very uncivill Answer to certain Observations upon Anthroposophia Theomagica, and Anima Magica Abscondita. Pp. 208.
Cambridge, by the printers to the University of Cambridge
1651
8° [12°, Wing]
Wing M2677; BL 526. f. 14. (4.)

Mystery of Alchimists (see ASHMOLE, *TCB*)

267 N., N.
**America: or an exact description of the West-Indies.*
London, by Ric. Hodgkinsonne for Edw. Dod
1655
8°
Wing N26; BL Thomason Collection

268 **Natura Exenterata: or nature unbowelled.*
London, printed for and sold by H. Twiford, G. Bedell and N. Elkins
1655
8°
Wing N241; BL Thomason Collection

269 NAUDÉ, Gabriel [NAUDEUS]
The History of Magick, by way of apology for all the wise men who have

unjustly been reputed magicians, from the Creation to the present age . . . Englished by J. Davies.
London, for John Streater
1657
8°
Wing N246; BL E. 1609. (1.)

270 NEDHAM, Marchamont
Medela Medicinae. A plea for the free profession and a renovation of the art of physick; out of the noblest . . . writers.
London, for Richard Lownds
1665
8°
Wing N397; BL 1038. f. 24.

271 NERI, Antonio
The Art of Glass, wherein are shown the wayes to make and colour glass, pastes, enamels, lakes, and other curiosities; . . . translated into English . . . To which is added an account of the glass drops made by the Royal Society.
London
1662: by A. W. for Octavian Pulleyn, 8° Wing N438; BL 1043. c. 20.
1672: not listed in Wing or BL

272 NICOLS, Thomas
A Lapidary: or, the History of pretious stones: with cautions for the undeceiving of all those that deal with pretious stones. Pp. 239.
1652: Cambridge, by T. Buck, 4° Wing N1145; BL 234. d. 32.
1653: [reissue, *Arcula Gemmea: or, a Cabinet of jewels. Discovering the nature, vertue, value of pretious stones*], London, for N. Brooke, 4° Wing N1143; BL 953. a. 11.
1659: *Gemmarius Fidelius, or the Faithful Lapidary*, London, H. Marsh, 4° Wing N1144; BL 444. c. 49.

273 NOLLIUS, Henry [NOLLE, Heinrich]
Chymists Key, or Doctrine of Corruption and Generation.
London, by E. B. for L. Lloyd
1657
8° & 16° [12°, Wing]
Wing N1221

274 NOLLIUS, Henry [NOLLE, Heinrich]

Hermetical Physick: or, the right way to preserve and to restore health . . . Englished by Henry Vaughan.
London, for Humphrey Moseley
1655
12°
Wing N1222; BL E. 1714. (1.)

275 NORMAN, Robert
The newe Attractive, containyng a short discourse of the Magnes or Lodestone; . . . concerning the Declinyng of the Needle, touched therewith under the plaine of the Horizon . . . Hereunto are annexed, certaine necessarie rules for the art of Navigation by the same R. N. 2 pt.
London, E. Allde, for H. Astley
1596
4° [8°, BL]
STC 18650; BL 51. c. 7.

NORTON, Thomas
Ordinall of Alchimie (see ASHMOLE, *TCB*)

NUISEMENT (see HESTEAU, Clovis)

276 O'DOWDE, Thomas.
The poor man's Physician, or the true Art of Medicine as it is Chymically prepared and administered, for healing the several diseases incident to mankind. [third edition.]
London, for F. Smith
1665
8°
Wing O139A; BL C. 124. c. 12.

277 OLDENBURG, Henry, Esq.
Philosophical Transactions, giving some accompt of the present undertakings, studies and labors of the ingenious in many considerable parts of the world [afterwards: *Philosophical Transactions of the Royal Society.*] *Begun as a periodical publication by H. Oldenburg, ... For the years 1665–70.*
London
1665
4°
BL L. R. 292.

278 P., H.
Five Treatises of the Philosophers Stone. Two of Alphonso, King of Portugall, as it was written with his own hand, . . . translated out of the Portuguez into English. One of J. Sawter . . . Another by R. Raudorff . . . Also a treatise of the names of the philosopher's stone, by W. Gratacolle [sic]. Translated into English. To which is added the Smaragdine Table. By the paines and care of H. P.
London, by Thomas Harper, to be sold by John Collins
1652
4°
Wing A2900; BL E. 654. (5.)

279 PARACELSUS
Paracelsus his Archidoxes, disclosing the genuine way of making Quintessences, Arcanums, Magisteries, Elixirs, &c.; Together with his books of Renovation & Restauration. Of the Tincture of the Philosophers. Of the Manual of the Philosophical Medicinal Stone. Of the Virtues of the Members. Of the Three Principles. And finally his seven books, of the Degrees and Composition of Receipts, and Natural Things. Faithfully and plainly Englished, and published by J. H. Oxon. 2 pt.
1660: London, for W. S. to be sold by T. Brewster, 24° Wing B3538
1661: London, for W. S., sold by Samuel Thomson, 8° Wing B3539; BL E. 2268.
1663: London, for Lodowick Lloyd, 8° Wing B3539A; BL 1489. k. 22.

280 PARACELSUS
Paracelsus his Aurora & Treasure of the Philosophers. As also the Water-stone of the Wise Men [by Ambrosius Siebmacher]; *describing the matter of, and manner how to attain the universal tincture. Faithfully Englished. And published by J. H. Oxon.* Pp. 229.
London, for Giles Calvert
1659
12°
Wing B3540; BL 8907. a. 22.

281 PARACELSUS
Paracelsus of the Chymical Transmutation, Genealogy and Generation of Metals & Minerals. With the Vertues and use of Dr. Trig's water, with the Mumial Treatise of Tentzelius. Also of the Urim and Thumim of the Jews . . . Whereunto is added, The philosophical and chymical experiments of Raymund Lully, with the right and due preparation of both Elixirs, and the perfect way of making the great Stone of Philosophers, as it was truly taught in Paris, and some time practiced in England by

R. [L]ully, in the time of King Edward the Third. Translated into English by R. Turner. Pp. 166.
London, for Rich. Moon & Henry Fletcher
1657
8°
Wing B3543; BL E. 1590. (3.)

282 PARACELSUS
Paracelsus his Dispensatory and Chirurgery, containing the choicest of his Physical Remedies . . . Faithfully Englished by W. D. Pp. 407.
London, T. M. for Philip Chetwind
1656
12°
Wing B3541; BL E. 1628.

PARACELSUS
An Excellent Treatise teaching howe to cure the French-Pockes (see HERMANNI, Philippus)

283 PARACELSUS
A hundred and foureteene Experiments and Cures of the famous Physitian Philippus Aureolus Theophrastus Paracelsus; Translated out of the Germane tongue into the Latin. Whereunto is added certaine excellent and profitable workes by B. G. a Portu Aquitano. Also certaine secrets of Isacke Hollandus concerning the vegetall and animall worke. Also the spagericke antidotarie for gunne-shot of Joseph Quirsitanus. Collected by John Hester. Pp. 84.
London, printed by Vallentine Sims
1596
4°
STC 19180; BL 778. e. 41. (1.)

284 PARACELSUS
The first part of the Key of Philosophie. Wherein is contained . . . secretes of Phisicke and Philosophie, divided into two bookes. In the first is the order to distill, oiles. In the seconde is the order to prepare mineralles. First written . . . by . . . Paraselius [sic], *and now published . . . by J. Hester.* 2 pt.
London
1580: R. Day, 8° STC 19181. 5
1596: Valentine Simmes, 8° STC 19181. 7; BL 1400. a. 34.
1633: [includes *A Store-house of physicall and philosophicall secrets*], A. M[athewes] for W. Lugger, 12° STC 19182; BL C. 118. a. 4.

PARACELSUS
His Mystical Philosophy written to the Athenians (see CROLLIUS, Oswald, *Admonitory Preface*)

PARACELSUS
Of the Nature of Things, 9 Books (see SENDIVOGIUS, Michael, *New Light of Alchymie*)

PARACELSUS
Philosophical and Chymical Treatise of Fire and Salt (possibly the *Philosophical and Chymical Experiments* by LULL, in *PARACELSUS, Chymical Transmutation*)

PARACELSUS
Secrets of Alchimy (see PARACELSUS, *Of the Supreme Mysteries of Nature*)

PARACELSUS
Secretes of Phisicke and Philosophie, &c. (see PARACELSUS, *Key of Philosophie*)

PARACELSUS
The Store-house of Physicall and Philosophical Secrets, in two parts (see PARACELSUS, *Key of Philosophie*)

285 PARACELSUS
Paracelsus of the Supreme Mysteries of Nature. Of the spirits of the planets. Occult philosophy. The magical, sympathetical and antipathetical cure of wounds and diseases. The mysteries of the twelve signs of the zodiack. Englished by R. Turner . . . Pp. 158.
London, J. C. for N. Brook and J. Harison
1656
8°
Wing B3544; BL E. 1567. (2.)

[In his *Continuation*, Cooper lists a 1659 edition of this work under the title, *Secrets of Alchimy*; this edition is not included in Wing or BL.]

PARACELSUS
Wonderfull Mysteries of the Creation (see CROLLIUS, Oswald, *Admonitory Preface*)

286 PARÉ, Ambroise
**The Works of that most famous chirurgion Ambrose Parey. Translated out of Latine and compared with the French. By Th: Johnson* [with woodcuts]. Pp. 1173.
London, Th: Cotes and R. Young
1634
fol.
STC 19190; BL C. 112. h. 9.

[Cooper's specific reference is to PARÉ's *Treatise of Distillations*, with Figures, being the 28th Book of his *Chirurgery*.]

287 PARTLITZ, Simeon von Spitzberg
New Method of Physick; or, a short view of Paracelsus and Galens Practise: in 3 treatises . . . Written in Latin by S. Partlicius . . . Translated into English by N. Culpeper.
London, by Peter Cole, to be sold in his shop and by S. Howes, J. Garfield, and R. Westbrook
1654
8°
Wing P612; BL E. 1475. (3.)

288 PARTRIDGE, John
Treasury of commodious conceits, & hidden secrets. Commonlie called, the good huswives closet.
London, R. Jones
1591
8°
STC 19429

PEARCE, the Black Monk (see ASHMOLE, *TCB*)

PENOTUS, Bernard G. a Portu Aquitano
Excellent Works (see FIORAVANTI, *Three Exact Pieces* and PARACELSUS, *A Hundred and Foureteene Experiments*)

PENOTUS, Bernard G. a Portu Aquitano
157 Philosophic Rules or Canons (see HELMONT, Franciscus Mercurius van, *One Hundred Fifty Three Chymical Aphorisms*)

289 PERSON, David
Varieties, or, a Surveigh of rare and excellent matters . . . Wherein the principall Heads of diverse Sciences are illustrated. Digested into five

Books, etc.
London, R. Badger for T. Alchorn
1635
4°
STC 19781; BL 537. b. 2.

[Cooper specifically cites Person's *Salamandra, or a short Treatise of the Philosophers Stone* in the 5th book of this work, p. 33.]

290 PETER, John
******A Treatise of Lewisham* (but vulgarly miscalled Dulwich) *Wells.*
London, by Tho. James for Sam. Tidmarsh
1680
12°
Wing P1691; BL 1171. e. 20. (1.)

PETTUS, Sir John
**Essays on Metallick Words* (see ERCKER, Lazarus)

291 PETTUS, Sir John
Fodinae Regales: or the history, laws and places of the chief mines and mineral works in England, Wales, and the English Pale in Ireland, as also of the mint and money . . . With a Clavis explaining some difficult Words relating to mines, &c. Pp. 108.
London, H. L. & R. B. for T. Basset
1670
fol.
Wing P1908; BL 434. e. 23.

292 PHAEDRO, George
Physicall and Chymicall Works to cure most difficile Diseases, with the secrets of Coelestial Physick. Selected out of the Germane and Latine language, by the industry of J. A. Schenccius. (Translated by N. Culpeper.)
London
1654: for William Sheares, 8° Wing P1955; BL E. 1497. (2.)
*1674: [third edition, retitled *The Art of Chymistry*], for Simon Neale, 8° Wing P1954; BL 1651/1423.

[**293 PHILALETHA, Eyrenaeus**, pseud.]
Secrets Reveal'd: or, an Open Entrance to the Shut Palace of the King; Containing, The greatest Treasure in Chymistry never yet so plainly discovered. Composed by a most famous English-man, Styling himself Anonymous, or Eyraeneus Philaletha Cosmopolita . . . Published . . . by

W. C., Esq., a true lover of Art and Nature [with the Epistle Dedicatory of John Langius, Publisher of this author in Latin and the "Preface of John Langius to the Reader"]. Pp. 120.
London, Wm. Godbid for William Cooper
1669
8°
Wing S5288; BL 8905. aa. 19.

PHILALETHES, Alazonomastix (see MORE, Henry)

PHILALETHES, Eirenaeus
**The Arcanum, or Secret of the Immortal Liquor Alkahest*
(see COOPER, William, *Collectanea Chymica*)

[**294 PHILALETHES, Eirenaeus Philoponos, pseud.**]
The Marrow of Alchemy, being an experimental treatise [in verse], *discovering the . . . Mystery of the Philosophers Elixer, in Two Parts. By Eirenaeus Philoponos Philalethes.*
London
1654: [part one], by A. M. for Edw. Brewster, 8° Wing S5278; BL C. 117. a. 16.
1655: [part two], by R. I. for Edw. Brewster, 8° Wing S5279; BL C. 117. a. 16.

[In his "Advertisement" prefacing the 1675 edition, Cooper states that "In the First Part of the Catalogue [i.e., 1673] . . . Geo. Starkie was put down for the Author of the *Marrow of Alchymie*, whereas he was but the Publisher."]

[**295 PHILALETHES, Eirenaeus, pseud.**]
**Ripley Reviv'd: or, an Exposition upon Sir George Ripley's Hermetico-Poetical Works. Containing the plainest and most excellent Discoveries of the most hidden Secrets of the Ancient Philosophers, that were ever yet Published. Written by Eirenaeus Philalethes an Englishman, stiling himself Citizen of the World* [Edited by William Cooper].
London, printed by Tho. Ratcliff and Nat. Thompson, for William Cooper
1678
8°
Wing S5286; BL 1034. f. 14.

["At the end Cooper has added in Latin: Porta Prima. De Calcinatione Philosophica; a chapter belonging to Philaletha's Fons Chymicae

Philosophiae," Ferguson, II, 193.]

Contents:
1. *An exposition upon Ripley's Epistle to King Edward IV.*
2. *An exposition upon Ripley's Preface.*
3. *An exposition upon Ripley's First six Gates of the Compound of Alchymie.*
4. *Experiments for the preparation of Sophick Mercury.*
5. *A Breviary of Alchemy.*
6. *An exposition upon Ripley's vision.*

PHILALETHES, Eirenaeus
Sir George Ripley's Epistle to King Edward IV unfoulded (see HARTLIB, Samuel, *CMCA*)

[**PHILALETHES, Eugenius**, pseud.] (see VAUGHAN, Thomas)

[**PHILANDER, Eugenius**] **GUIDOTT, Thomas**
Quere concerning drinking of Bath-water at Bath, resolved (see JORDEN, Edward, *Discourse of Naturall Bathes* [fourth edition])

Philosophia Maturata (see COELSON, Lancelot)

A Philosophical Riddle of the Agreement of Metals (see HOUPREGHT, *Aurifontina Chymica*)

**A Physical Dictionary* (see BLANKAART, Stephen)

296 *A Physical Dictionary, or an Interpretation of such crabbed Words and Terms of Art, as are derived from Greek and Latin, used in Physick, Anatomy, Chyrurgery and Chimistry* (see also RENODAEUS, *Medicinal Dispensatory*).
London, by G. D. for John Garfield
1657
8° [12°, Wing]
Wing P2143

[Cooper adds "published by Garfield for Explanation of Rich. Tomlin's Translation of Renodaeus's *Dispensatory.*"]

297 PLATT, Sir Hugh
**Floraes Paradise, beautified and adorned with sundry sorts of delicate fruites and flowers . . . With an offer of an English Antidote.*

London, H. L[ownes] for W. Leake
1608
8° [12°, BL]
STC 19990; BL C. 123. b. 17.

[Cooper adds "This Book is but the first Edition, and first part of the Authors Garden of Eden: But there is this difference that the new invention of Light Plate and English Wines is left out of the later Editions," *Continuation*, D* 19]

298 PLATT, Sir Hugh
The Garden of Eden. Or, an Accurate Description of all Flowers and Fruits now growing in England, with particular rules how to advance their nature and growth, as well in Seeds as Herbs, and the secret ordering of Trees and Plants; Part I (two editions); Part II; and all printed together.
London
1653 & 1655: W. Leake, 8° Wing P2385 & P2386A; BL F. 1294. (1.)
1660: [second part], W. Leake, 8° Wing P2392; BL E. 1804. (2.)
1674 [1675, BL and Wing]: [all together], for Wm. & John Leake, 8° Wing P2388; BL 449. a. 25. (2.)

299 PLATT, Sir Hugh
The Jewell House of Art and Nature. Conteining divers Chymicall Conclusions of the Art of Distillation, &c. . . . Inventions, together with . . . experimentes in the art of Husbandry, Distillation, and Moulding.
3pt.
London
1594: P. Short, 4° STC 19991; BL 959. a. 4.
*1653: [with *A Rare and excellent discourse of minerals, stones, gums, and resins . . . by D. B., Gent.*], Bernard Alsop, 4° Wing P2390; BL C. 136. f. 26.

300 PLATT, Sir Hugh
**Sundrie new and artificiall Remedies against Famine . . . uppon thoccasion [sic] of the present dearth.*
(see also COOPER, *Collectanea Chymica*)
London, P. S[hort]
1596
4°
STC 19996; BL C. 31. c. 10.

PLATTES, Gabriel

Caveat for Alchymists to prevent cheating (see HARTLIB, *CMCA*)

301 PLATTES, Gabriel
A Discovery of Infinite Treasure hidden since the Worlds beginning. Whereunto all men, of what degree soever, are friendly invited to be sharers with the discoverer, G. P. Pp. 92.
London, J. Leggatt for G. Hutton
1639
4°
STC 19998; BL 966. b. 31.

302 PLATTES, Gabriel
A Discovery of Subterraneall Treasure: viz. of all manner of Mines and Mineralls from the Gold to the Coale, with plain directions and rules for the finding of them, in all kingdoms and countries; and also the art of melting, refining, and assaying of them is plainly declared, how any ordinary man may try the value of Oares found either by Rule or Accident.
London
1639: J. Okes for J. Emery, 4° STC 20000; BL 727. c. 15.
*1653: [second edition], J. E., to be sold by Humphrey Moseley, 4° Wing P2410; BL E. 686. (11.)
*1679: [third edition], for Peter Parker, 4° Wing P2411; BL B. 394. (9.)

[Cooper mistakenly places this entry under Platt, Sir Hugh; he further reports that *A Discovery* is included in Moses Cook, *The Manner of Raising, Ordering, and Improving Forrest-trees*, London, 1679, 4° Wing C6033.]

303 PLINIUS SECUNDUS, Caius
Secretes and wonders of the worlde, containing many wonderfull properties given to Man, Beasts, Fowles, Fishes, and Serpents, Trees, and Plants, abstracted out of . . . Plinie. Translated out of French into English.
London, printed for Tho. Hackett
1587
4°
STC 20033; BL 157/766.

304 PLOT, Robert
**The Natural History of Oxfordshire, being an essay towards the natural history of England, with a map of Oxfordshire, and 16 Copper Figures.* Pp. 358.

Oxford, printed at the Theater in Oxford, and are to be had there: and in London at Mr. Moses Pits, and at Mr. S. Millers
1676 [1677, BL]
fol.
Wing P2585; BL 10351. h. 12.

305 POLEMAN, Joachimus
Novum Lumen Medicum; wherein the excellent and most necessary doctrine of the highly gifted philosopher Helmont concerning the great mystery of the Philosophers sulphur is fundamentally cleared. Written by the authour in . . . German, and Englished by F. H., a German.
London, printed by J. C. for J. Crook
1662
8°
Wing P2748; BL 1651/1309.

PONTANUS, Johann
Epistle upon the Minerall Fire (see FLAMEL, Nicolas, *Hieroglyphicall Figures*, and THOR, George, *Cheiragogia Heliana*)

306 PORTA, Giovanni Battista della
Natural Magick . . . in twenty Books; wherein are set forth all the Riches and Delights of the natural Sciences. [Translated from the Latin.] Pp. 409.
London, T. Young and S. Speed
1658
fol. [4°, BL]
Wing P2982; BL C. 48. h. 12.

307 POWELL, Thomas
Humane Industry: or, the History of most Manual Arts, deducing the original, progress and Improvement of them; with variety of instances and examples
London, for Henry Herringman
1661
8°
Wing P3072; BL 1042. c. 2.

308 POWER, Henry
Experimental Philosophy, in three books, containing new experiments–microscopical, mercurial, magnetical—with some deductions and probable hypotheses raised from them, in avouchment and illustration of the

now famous atomical hypothesis.
London, by T. Roycroft, for John Martin, and James Allestry
1664
4°
Wing P3099; BL 537. h. 1.

**The Practice of Lights* (see COOPER, William, *Collectanea Chymica*)

309 PRAEPOSITI, Nicolaus
Prepositas his Practise, a worke . . . for the better preservation of the Health of Man. Wherein are . . . approved Medicines, precious Waters, Receipts and Ointmentes . . . Translated out of the Latin into English by L[eonard?] M[ascall?]. Pp. 111.
London, J. Wolfe for E. White
1588
4°
STC 20180. 7; BL 1038. k. 47. (1.)

310 PRIMROSE, James
**Popular Errours, or the errours of people in Physick; first written in Latine by . . . J. Primrose . . . Translated into English by Robert Wittie* [Witty].
London, by W. Wilson for Nicholas Bourne
1651
8°
Wing P3476; BL E. 1227.

311 **The Principles of the Chymists of London stated, with the Reasons of their Dissent from the Colledge of Physitians, as they were unanimously agreed on by them . . . and are now published for the Information of the said Colledge, . . . as also for instructing the People in what is most useful for their Preservation.* 2 pt. Pp. 128.
London, for the authors
1676
8°
Wing P3497; BL 778. c. 38.

The Privy Seal of Secrets . . . of the Philosophers
(see HOUPREGHT, *Aurifontina Chymica*)

A Profitable boke . . . (see MASCALL, Leonard)

312 ******A Profitable discourse Composed by divers Grocers against the bad Garbelling of Spices used in these days, and against the Combination of the Workmen of that office.*
London, printed for Tho. Man
1591
4°

[Entry found only in Maunsell.]

A Prognostication of what shall happen to Physitians, Chirurgians, Apothecaries, and their dependents, with Alchymists and Miners
(see HARTLIB, *CMCA*)

313 PRYNNE, William
Aurum Reginae; or, a compendious tractate, and chronological collection of Records in the Tower and Court of Exchequer concerning Queen-Gold . . . Four Patents granted by King Henry the Sixth to certain Persons, by Authority of Parliament for finding the Philosophers-Stone, to transubstantiate baser metals into solid real Gold and Silver, to satisfie the Creditors of the King, and Kingdom. 2 pt.
London, for the author by T. Ratcliffe, to be sold by Edward Thomas, and Joshua Robinson
1668
4°
Wing P3898; BL 1129. k. 3. (1, 2.)

314 *Public Bathes purged, or a Reply to Dr. Chamberlens Vindication of Public Artificial Bathes from the pretended Objections and Scandals obtruded on them.* (see CHAMBERLEN, Peter, *Vindication*, and ALSTON, Sir Edward, *A Paper delivered in . . .*)
n.p.
1648
4°
Wing C1902 [but entry cancelled in Wing, second edition.]

[This entry may be the same as, or closely related to, Sir Edward Alston, *A Paper delivered in. . . .*]

PYTHAGORAS
Wheele of Fortune (see CATTAN, Christophe, *Three Books of Geomancy*)

[**QUERCETANUS**] (see DUCHESNE, Joseph)

RANDORFF, Florianus
Of the Philosophers Stone (see P., H., *Five Treatises*)

315 RAY, John
**A Collection of English Words not generally used, with their significations and original, with catalogues of English birds and fishes; and an account of the preparing and refining such metals as are gotten in England.* Pp. 178.
London, H. Bruges, for T. Burrell
1674
8°
Wing R388; BL 435. a. 26.

316 RAYNALDE, Thomas
A compendious Declaration of the excellent vertues of a certain lateli inventid oile, callid . . . Oile Imperial. With the maner how the same is to be usid, etc. [Translated from A. Vesalius' *Radicis chynae usus.*]
London [Venetiis, STC, BL], J. Gryphius
1551
8°
STC 20779; BL 1400. a. 33.

REDMAN, William
Aenigma Philosophicum (see ASHMOLE, *TCB*)

317 *A Relation concerning a wonderful Fountain in Germany near the City Halberstad.*
London
1646
4°
Wing R773

RENODAEUS, Joannes [J. de RENOU]
Conference concerning the Phylosophers Stone (see HARTLIB, *CMCA*)

RENODAEUS, Joannes [J. de RENOU]
Conference, whether or no each several Disease hath a particular Remedy (see HARTLIB, *CMCA*)

318 RENODAEUS, Joannes [J. de RENOU]
A Medicinal Dispensatory, containing the whole body of physick; discovering the natures . . . of vegetables, minerals and animals, the man-

ner of compounding medicaments: methodically digested in five books of philosophical and pharmaceutical Institutions; three books of physical materials . . . Together with a . . . Pharmacopoea . . . now Englished and revised by R. Tomlinson.
London, by Jo. Streter and Ja: Cottrel
1657
fol. [4°, BL]
Wing R1037; BL 546. i. 6.

REYNOLDS, John
**The useful and easie tables . . .* (see BADCOCK, William, *A touch-stone*)

RHUMELIUS, Johann Conrad
Experiments (see COELSON)

319 **RIDLEY, Mark**
**Magneticall Animadversions. . . . upon certaine Magneticall Advertisements lately published, from Maister W. Barlow.*
London, N. Okes
1617
4°
STC 21044; BL 538. f. 42.

320 **RIDLEY, Mark**
Short Treatise of magnetical bodies and motions.
London, N. Okes
1613
4°
STC 21045; BL C. 112. b. 10.

RIPLEY, Sir George
Bosome Book (see COOPER, *Collectanea Chymica*)

RIPLEY, Sir George
Compound of Alchimy; on the Emblematical Scrowl; Epistle unto King Edward the Fourth; Mystery of the Alchymists; Preface to Medulla; Short Work; Twelve Gates; his Verses; Vision; Wheel (see ASHMOLE, *TCB*)

321 **RIPLEY, Sir George**
The Compound of Alchymy, or the ancient hidden Art of Alchemie: conteining the right and perfect means to make the Philosophers Stone, Au-

rum Potabile; with other excellent Experiments . . . first written by G. Ripley . . . Whereunto is adjoyned his Epistle to the King . . . & other his workes never before published. With certaine brief additions . . . set foorth by R. Rabbards, etc. [in verse] (see also ASHMOLE, *TCB*)
London, T. Orwin
1591
4°
STC 21057; BL C. 39. d. 26.

RIPLEY, Sir George
Treatise of Mercury, and the Philosophers stone (see HOUPREGHT, *Aurifontina Chymica*)

322 RIVERIUS, Lazarus
**The Secrets of the famous L. Riverius . . . Newly translated from the Latin, by E. P., M. D.*
London, for Daniel Brown
1685
8°
Wing R1565; BL 7461. a. 59.

323 ROBINSON, John, M. D.,
**Endoxa, or, some probable inquiries into Truth, both divine and humane: together with a Stone to the Altar; . . . as also, a calm ventilation of Pseudodoxia Epidemica (or, Doctrine of Vulgar Errors, set forth . . . by Thomas Browne).* 3 pts.
London, J. Streater, for F. Tyton
1658
8°
Wing R1700; BL E. 1821. (1.)

ROBINSON, Thomas
Of the Philosophers Stone (see ASHMOLE, *TCB*)

324 RODOCANACES, Constantinos [RHODOKANAKIS, Konstantinos]
Alexicacus: Spirit of Salt of the World, which vulgarly prepared, is called the Spirit of Salt, . . . or the transcendent virtue of the true Spirit of Salt, now philosophically prepared and purified from all hurtful and corroding qualities.
London
*1664: [third edition], by R. D., 4° Wing R 1332; BL 778. e. 41. (6.)

1670: [seventh edition, enlarged], by W. G., 4° Wing R1333; BL 778. e. 41. (10.)

325 RODOCANACES, Constantinos [RHODOKANAKIS, Konstantinos]
**A Discourse in the praise of antimonie and the vertues thereof.*
London
1664
4°
Wing R1335; BL 7461. aaa. 63.

326 ROVENZON, John
Treatise of Metallica, but not that which was published by Mr Simon Sturtevant upon his patent, which is now by order cancelled . . . Whereupon priviledge is granted . . . to John Rovenzon, Esquire, for the making of Iron and other Metals and Materials with Sea-Coale, Pit-Coale, &c. with a brief Explanation, Demonstration and Discovery of the Inventions, Instruments, Engines, Furnaces, &c. with the materials, things and works to be made by said Fewel. (see STURTEVANT, Simon)
London, N. Okes for T. Thorpe
1673 [1613, STC and BL]
4°
STC 21355; BL 727. c. 14.

327 ROWZEE, Lodowick
The Queenes Welles, that is, a treatise of the nature and vertues of Tunbridge water, etc. Pp. 79. [editions of 1632 work]
London, for R. Boulter
1670: 8° Wing R2098; BL G. 3104.
1671: 8° Wing R2099; BL 1171. e. 33.

328 RUSSEL, Thomas
Diacatholicon Aureum; or, a generall powder of Gold, purging all offensive humours in Mans body: good in generall for all diseases, where there needeth any Purgation, as also for any sicknesse yearely incident, etc.
London, for J. Flasket
1602
4°
STC 21455; BL 1034. h. 18. (1.)

329 RUSSELL, William
**A Physical Treatise, grounded not upon tradition nor phancy, but expe-*

rience. Pp. 179.
London, J. Williams
1684
8°
Wing R2357; BL 775. b. 27.

SALA, Angelo
Angelo Sala's Preparation (see COELSON, *Philosophia Maturata*)

330 SALMON, William
**Polygraphice*; or the Arts of drawing, engraving, . . . varnishing, . . . and dying, etc. [fifth edition, enlarged].
London, for T. Passinger and T. Sawbridge
1685
8°
Wing S448; BL 1043. d. (9.)

[In this volume, Cooper specifically refers to Peter John Faber's *112 Chymical Arcanums*, "being the 4th. and last Book of his *Universalis sapientia*."]

331 SALMON, William
Synopsis medicinae or compendium of Astrological, Galenical and Chymical physick and philosophy, deducted from the Principles of Hermes and Hippocrates. In three books, etc.
London, by W. Godbid, for Richard Jones
1671
8°
Wing S454; BL 774. f. 2.

332 SANDERS [SAUNDERS], Richard
**Physiognomie, and Chiromancie, Metoposcopie the symmetrical proportions and signal moles of the body, fully and accurately handled: with their natural-predictive significations. The subject of dreams . . . Whereunto is added the Art of Memorie* (or, the art of R. Lullius).
London
1653: R. White for N. Brook, fol. Wing S754; BL 719. k. 7.
1671: [second edition], by H. Brugis for Nathaniel Brook, fol. Wing S755; BL G. 2517.

SAWTRE, John
Of the Philosophers Stone (see P., H., *Five Treatises*)

333 SCHROEDER, Johann
The Compleat Chymical Dispensatory in five books: treating of all sorts of metals, precious stones, minerals, vegitables, and animals; and how rightly to know and use them . . . Englished by W. Rowland.
London, by John Darby, for Richard Chiswell, and Robert Clavell
1669
fol.
Wing S898; BL 7510. g. 21.

334 SCHROEDER, Johann
Ζωολογια : or the History of Animals, as they are useful in physick and chirurgery . . . Translated by T. B., i.e., T. Bateson.
London, T. Cotes for R. Royston
1669[1659, Wing, BL]
8°
Wing S899; BL E. 1759. (1.)

335 SCOT, Patrick
The Tillage of Light, or, a true discoverie of the philosophicall Elixir, commonly called the Philosopher's Stone.
London, for W. Lee
1623
8°
STC 21862; BL 1033. d. 9.

336 SCRIBONIUS, Gulielmus Adolphus
Naturall Philosophy: or Description of the World and several Creatures . . . therein, (viz.) *of Angels, Mankind, Heaven, Stars, Planets, the 4 Elements with their order, nature and government, as also of minerals, Metals, plants, and pretious stones, with their Colours, Forms and Vertues* (translated by Daniel Widdowes). [second edition].
London, T. Cotes, for J. Bellamie
1631
4°
STC 22112; BL B. 745. (5.)

Seal of Secrets (see HOUPREGHT, *Aurifontina Chymica*)

Secretes and wonders of the worlde (see PLINIUS SECUNDUS)

Secrets disclos'd of the Philosophers Stone (see HOUPREGHT, *Aurifontina Chymica*)

337 *Secrets for liquifying and using of Gold and Silver.*
London
8°

[No additional information found, but possibly an edition of *Booke of Secrets.*]

SEILERUS, Wenceslaus (see BECHER, Johann)

338 SENDIVOGIUS, Michael
A New Light of Alchymie: taken out of the fountaine of nature, and manuall experience. To which is added a treatise of sulphur . . . by M. Sandivogius [sic]. *Also nine books Of the Nature of Things, written by Paracelsus . . . Also a Chymicall dictionary explaining hard places and words in the writings of Paracelsus and other obscure authors . . . Translated out of the Latin . . . by J. F., M. D.*
London
1650: R. Cotes, for Thomas Williams, 4° Wing S2506; BL E. 604. (3.)
*1674: for Thomas Williams, 12° [8°, BL] Wing S2507; BL 1484. ee. 8.

339 SENNERTUS, Daniel
**Chymistry made Easie and Useful. Or, the Agreement and Disagreement of the Chymists and Galenists. By D. Sennertus, N. Culpeper and Abdiah Cole.* Pp. 166.
London, P. Cole
1662
8°
BL 7580. a. 49. (2.)

340 SENNERTUS, Daniel
The Institutions or Fundamentals of the whole Art, both of Physick and Chirurgery, divided into five books . . . wherein are the grounds, of Chymistry . . . made English by N. D. B. P., late of Trinity College in Cambridge. Pp. 492.
London, L. Lloyd
[1656]
fol. and 8°
Wing S2535; BL E. 1568.

[Cooper shows editions in fol. and 8° but records no dates.]

341 SENNERTUS, Daniel
**The Sixth Book of Practical Physick. Of Occult or Hidden Diseases*

in Nine Parts; . . . By D. Sennertus, N. Culpeper and Abdiah Cole [or rather, translated by Culpeper and Cole from Sennertus.] [With "Two Epistles by Balthasar Han."]
London, Peter Cole
1662
8°
Wing S2539/40 [?]; BL 7580. a. 49. (3.)

342 SENNERTUS, Daniel
*[*T*]hirteen Books of Natural Philosophy. . . . Eight Books of Natural Philosophy . . . Unto which is added five books more of natural philosophy in several discourses . . . Written in Latin and English. By Daniel Sennert . . . Nicolas Culpeper . . . Abdiah Cole* [or rather, translated by Culpeper and Cole from Sennertus.] Pp. 530.
London, Peter Cole and Edward Cole
1661
fol.
Wing S2545; BL 773. m. 10. (2.)

343 SHARROCK, Robert
The History of the propagation and improvement of Vegetables, by the concurrence of art and nature, usually cultivated in England and according to Observations made from Experience and Practice.
Oxford
1660: by A. Lichfield for Tho. Robinson, 8° Wing S3010; BL E. 1731. (2.)
1671 [1672]: [Second edition, much enlarged], by W. Hall for R. Davis, 8° Wing S3011; BL 972. b. 6.

344 SHIRLEY, Thomas
A Philosophical Essay: declaring the probable causes, whence stones are produced in the greater world, with a search into the Origin of all bodies.
London, for William Cademan
1671 [1672, Wing, BL]
8°
Wing S3523; BL 954. a. 12.

Shop of five Windows (see *Key to unknowne Knowledge*)

SIEBMACHER, Johann Ambrosius
The Water-stone of the Wise Men, describing the matter of, and manner how to attain the universal tincture (see PARACELSUS, *Aurora*)

345 SIMPSON, William
**The History of Scarborough Spaw.*
London, for Thomas Simmons
1679
8° [12°, Wing]
Wing S3832; BL 1171. e. 44.

346 SIMPSON, William
Hydrologia Chymica: or the Chymical Anatomy of the Scarbrough, and other Spaws in Yorkshire. Wherein are interspersed some animadversions upon Dr. Witties Treatise of Scarbroughs Spaw, with a Description of the Spaws at Malton and Knarsbrough, with the Original of Springs, Fountains, &c.
London, by W. G. for Richard Chiswell
1669
8°
Wing S3833; BL 1171. e. 18. (1.)

347 SIMPSON, William
Hydrological Essayes: or, a Vindication of Hydrologia Chymica, being a further discovery of the Scarborough Spaw, and of the right use thereof . . . with an account of Allom-works at Whitby, &c. To which is annexed, an answer to Dr. Tunstal's book concerning Scarborough Spaw.
London, J. D. for Richard Chiswell
1670
8°
Wing S3834; BL 1171. e. 45.

348 SIMPSON, William
** Philosophical Dialogues concerning the Principles of Natural Bodies, wherein the principles of the old and new philosophy are stated, and the new demonstrated, more agreeable to reason, etc.*
London, by T. Hodgkin for Dorman Newman
1677
12°
Wing S3835; BL 536. a. 22.

349 SIMPSON, William
**Two small Treatises: the first, a further essay towards the history of this present Fever with its two products the Morbus Cholera, and the Grypes; . . . the second, a medico-philosophical analogy betwixt vegetable and animal juyces.* 2 pt.
London, by T. H. for E. Wyer

1678
12° [8°, Wing]
Wing S3838; BL 1167. b. 9. (2.)

350 SIMPSON, William
Zymologia Physica, or a brief Philosophical Discourse of Fermentation, from a new Hypothesis of Acidum and Sulphur, whereby the Phoenomena of all natural Hot-Baths, the Generation of Minerals, the Production of many Acidulae or Spaw-waters, the grand appearances of Heat, Fire and Light throughout the triplicity of Natures Dominions, &c.; with a Discourse of the Sulphur-Bath at Knaresbrough in Yorkshire.
London, by T. R. & N. T. for W. Cooper
1675 [1670, BL]
8°
Wing S3840; BL 1170. d. 39.

351 SPRAT, Thomas
History of the Royal Society of London, for the improving of natural knowledge. Pp. 438.
London, T. R. for J. Martyn, and J. Allestry
1667
4°
Wing S5032; BL 740. c. 17.

352 STANHOPE, Michael
Cures without Care, Or a summons to all such who finde little or no helpe by the use of ordinary physicke to repaire to the Northerne Spaw. Wherein . . . it is evidenced to the world, that infirmities in their owne nature desperate and of long continuance have received perfect recovery, by vertue of Minerall waters neare Knaresborow . . . Also a description of the said water . . . Faithfully collected for the publique good; by M. S[tanhope].
London, W. Jones
1632
4°
STC 23226; BL 1171. d. 47.

353 STANHOPE, Michael
Newes out of York-shire: or, an account of a journey, in the true discovery of a soveraigne Minerall, Medicinal Water . . . neere . . . Knaresbrough, not inferiour to the Spa in Germany.
London, J. H. for G. Gibbes
1627

4°
STC 23228; BL G. 3115.

354 STANLEY, Thomas, translator
The Chaldaick Oracles of Zoroaster and his followers. With the Expositions of Pletho and Psellus [in *History of the Chaldaick Philosophy*, trans. Thomas Stanley].
London, for Thomas Dring
1662
fol.
Wing S5240; BL 523. k. 4. (2.); Dr. Williams's Library, 1011. s. 11.

355 STARKEY, George
Admirable Efficacy . . . of Oyl which is made of Sulphur-Vive, set on fire, and called commonly Oyl of Sulphur per Campanam . . . Faithfully collected out of the writings of John Baptiste van Helmont (see COOPER, *Collectanea Chymica*).
London
1660
8°

[BL and Wing list no 1660 edition as reported by Cooper.]

356 STARKEY, George
A brief Examination and Censure of several Medicines of late years extoll'd for Universal Remedies and Arcana's of the highest Preparation, being some of them trivial Toyes, by mistake cryed up for what they are not, others dangerous, virulent venomes, falsely commended for rare wonder-working Mysteries, from which injuries true Chymistry is vindicated, and the disguise pluck't off from counterfeit Artists.
London, for the Author
1664
8° [12°, Wing, BL]
Wing S5272; BL 1506/823.

357 STARKEY, George
**G. Starkey's Pill vindicated from the unlearned alchymist, and all other pretenders, etc.* (see also KENDALL, George and MATHEWS, Richard)
London
1684 [1660(?), BL and Wing]
4°
Wing S5283; BL 778. e. 41. (14.)

[Cooper lists only the edition included in *Collectanea Chymica*, 1684.]

358 STARKEY, George
Liquor Alchahest, or a Discourse of that Immortal Dissolvent of Paracelsus and Helmont, it being one of those two wonders of Art and Nature, which radically dissolves all Animals, Vegetals, and Minerals into their principles, without being in the least altered, either in weight or activity after a thousand dissolutions, &c. Pp. 55.
London, by T. R. & N. T., for W. Cademan
1675
8°
Wing S5277; BL 717. a. 50.

359 STARKEY, George
Nature's Explication and Helmont's Vindication: or a Short and sure Way to a long and sound life . . . being a full apology for chymical medicaments with a vindication of their Excellencies, against the Galenists.
London, by E. Cotes, for Thomas Alsop
1657[8]
8°
Wing S5280; BL E. 1635. (2.)

360 STARKEY, George
Pyrotechny asserted and illustrated, to be the surest and safest means for arts triumph over nature's infirmities, being a full and free discovery of the medicinal mysteries studiously concealed by all artists, and only discoverable by fire: With an Appendix concerning the nature, preparation, and virtue of the great Arcana.
London, by R. Daniel, for Samuel Thomson
1658
8°
Wing S5284; BL 1035. a. 13.

361 STARKEY, George
**A smart Scourge for a silly, sawcy Fool; being an answer to a letter, at the end of a pamphlet of L. Lockyer (quondam . . .) Botcher, now drest up with the title of licensed Physician.*
London
1664 [1665, Wing, BL]
4°
Wing S5289; BL 551. a. 57. (1.)

STARKEY, George
Unfolding of Sir George Ripleys Epistle to King Edward the Fourth, being his Key of Philosophy without contradictions, and the substance of Philosophy by Conclusions (see HARTLIB, *CMCA*)

Store-house of physicall and philosophicall secrets (see PARACELSUS, *Key of Philosophie*, 1633 edition)

362 STUBBE, Henry
**The Indian Nectar, or, a Discourse concerning Chocolate, etc.*
London, by J. C. for Andrew Crook
1662
8°
Wing S6049; BL 450. d. 9.

363 STUBBE, Henry
**Legends no Histories: or, a Specimen of some animadversions upon the "History of the Royal Society"* [by Dr. Sprat], *wherein besides the several Errors against Common Literature, sundry mistakes about the making of Salt-Petre and Gun-powder, are detected and rectified, . . . together with the "Plus Ultra" of Mr. Jos. Glanville reduced to a non-plus &c.*
London
1670
4°
Wing S6053; BL 740. c. 19. (1, 2.)

364 STUBBE, Henry
Lex Talionis; sive vindiciae pharmacoporum; or a short Reply to Dr. Merret's Book and others, written against the Apothecaries, discovering the frauds and abuses committed by Doctors professing and practising Pharmacy.
London, printed, and are to be sold by Moses Pitt
1670
4° [8°, Wing]
Wing S6055; T. C. I 38

365 STUBBE, Henry
Medice, cura teipsum. Or The apothecaries plea in some short and modest animadversions, upon a late tract entitled A short view of the frauds and abuses of the apothecaries, . . . by Christopher Merret . . . From a real well-wisher to both societies [i.e., Henry Stubbe].
London, for W. Miller

1671
4°
Wing S6061; BL 551. a. 44.

366 STURTEVANT, Simon
Metallica; or the Treatise of Metallica. Briefly comprehending the doctrine of diverse New Metallicall inventions, but especially how to neale, melt, and worke all kinde of mettal-oares, Irons and Steels with Sea-Coal, Pit-Coal, Earth-Coal, and Brush-Fewel . . . Also a transcript of his Majesties letters. . . . (see ROVENZON, John)
London, G. Eld.
1612
4°
STC 23411; BL 726. i. 28.

367 SUCHTEN, Alexander von
A. van S. of the secrets of antimony: in two treatises with Basil Valentine's Salt of Antimony, and its use. Translated out of High Dutch by Dr. C.
London, for Moses Pitt
1670
8° [12°, BL]
Wing S6117; BL 1033. f. 35.

SYNESIUS
Book concerning the Philosopher's Stone (see VALENTINE, Basil, *Triumphant Chariot of Antimony*, 1678 edition)

368 TACHENIUS, Otto
**O. Tachenius his Hippocrates chymicus, discovering the ancient foundation of the late Viperine salt with the Clavis thereunto annexed. Translated by J. W.* 2 pt.
London, printed to be sold by Nath. Crouch
1677
4°
Wing T98; T.C. I 266

369 TAISNIER, Joannes
**A very necessarie and profitable Booke concerning Navigation, compiled in Latin by Joannes Taisnierus . . . named a treatise of continuall Motions. Translated into Englishe, by Richarde Eden.*
London, R. Jugge

n.d. [1575?, STC; 1579?, BL]
4°
STC 23659; BL 51. c. 51.

TALBOT, Sir Gilbert
Powder of Sympathy (see HIGHMORE, Nathaniel, *History of Generation*)

370 TENTZEL, Andrea
Medicina Diastatica, or sympatheticall Mumie . . . containing many mysterious and hidden Secrets in Philosophy and Physick, teaching the Magnetical Cure of Diseases at distance, &c. abstracted from the works of Theoph. Paracelsus, by the labour and industry of Andrea Tentzelius, first part (see also PARACELSUS, *Chymical Transmutation*).
London
1653
8° [12°, BL]
Wing T728; BL 1036. a. 18.

TENTZEL, Andrea
The Second Part of his Treatise of Mumie (see PARACELSUS, *Chymical Transmutation*)

TERNE, Christopher (see CLERKE, Timothy)

371 *A theologicall and philosophical treatise of the nature and goodnesse of salt. Or a New Compost for barren Land; wherein the old vulgar conceit that Salt makes Earth barren is examined, and the more ancient Truth, that it is the fructifier of the Earth is established.*
London, F. Kyngston for R. Boyle
1612
4°
STC 24233; BL 7076. a. 56.

372 THEVET, André
**The New found worlde, or Antarctike, wherin is contained beastes, fishes, . . . written in French . . . And now newly translated into English* [by T. Hacket].
London, Henrie Bynneman, for Thomas Hacket
1568
4°
STC 23950; BL G. 7107.

373 THIBAUT, Pierre
The Art of Chymistry: as it is now practised . . . Translated into English by a Fellow of the Royal Society.
London, for John Starkey
1668: 8° Wing T891; BL 8906. de. 43.
1675: 8° [12°, Wing], Wing T892; BL 1486. aa. 4.

374 THOMPSON, James
Helmont Disguised; or the vulgar errours of impericall and unskillful Practisers of Phisick confuted . . . in a dialogue between Philatrus and Pyrosophilus.
London, by E. Alsop, for N. Brook, and W. Leybourn
1657
8°
Wing T999; BL 775. b. 19.

375 THOMSON, George
Αιματιασις; or the true way of preserving the Bloud in its integrity, and rectifying it, if at any time polluted and degenerate; discovering the Errours of bleeding of Dr. Willis and Dr. Betts . . . Whereunto are added a stomachical spirit . . . Diaeterical instructions, etc.
London, for Nat. Crouch
1670
8°
Wing T1021; BL 783. e. 12. (1.)

376 THOMSON, George
Galeno-pale, or a Chymical Trial of the Galenists, that their Dross in Physick may be discovered. With the Grand abuses and disrepute they have brought upon the whole art of Physick and Chirurgery.
London, by R. Wood, for Edward Thomas
1665
8°
Wing T1023; BL 1167. e. 47.

377 THOMSON, George
Letter sent to Mr. Henry Stubbe, wherein the Galenical Method & Medicaments, as likewise Blood-letting in particular are offered to be proved ineffectual, by Experimental Demonstrations. . . . Also his Answer, . . . on which animadversions are made by G. Thomson, etc.
London, for Nath. Crouch
1672
4°

Wing T1025; BL 551. a. 9. (5.)

378 THOMSON, George
Λοιμοτομια; or, the Pest anatomized, . . . Together with the authors apology against the calumnies of the Galenists, and a word to Mr. N. Hodges concerning his late Vindiciae Medicinae.
London, for Nath. Crouch
1666
8°
Wing T1027; BL 1167. c. 26.

379 THOMSON, George
μεθοδος ιατροχυμικη; or, the direct method of curing chymically. Wherein is conteined the original matter, and principal agent of all natural bodies, also the efficient and material cause of Diseases in general; . . . with the Art of Midwifery chymically asserted . . . A Description of the Sanative Virtues of our Stomach Essence etc.
London, for B. Billingsley, & S. Crouch
1675
8°
Wing T1029; BL 1163. b. 54.

380 THOMSON, George
Μισοχυμιας Ελεγχος; or, a check given to the insolent garrulity of H. Stubbe: in vindication of My Lord Bacon and the author, with an assertion of experimental philosophy. Also some practical observations exhibited for the credit of the true chymical science. Lastly a brief contest between the Thomsons and the Merrets, who are the best Physicians.
London, for Nat. Crouch
1671
8°
Wing T1028; BL 1167. e. 49.

381 THOMSON, George
Πλανο-Πνιγμος; or a Gag for Johnson that published Animadversions upon Galeno-pale, and a Scourge for that pitiful fellow Mr. Galen, that dictated to him a Scurrillous Greek Title . . . 2 pt.
London, by R. Wood, for Edward Thomas
1665
8°
Wing T1030; BL 1167. e. 48.

382 THOR [THORNLEY], George
Cheiragogia Heliana. A manuduction to the philosophers magical gold . . . . To which is added . . . Zoroaster's Cave; or, an intellectual echo, &c. Together with the famous Catholic Epistle of John Pontanus upon the minerall fire.
London
1659: for Humphrey Moseley, 8° Wing T1037; BL 1033. d. 30.
1667: for Matthew Smelt, 8° Wing T1038

[The *Continuation* contains the following note on this entry: "Geo. Thornley's twelve Eccho's said to be a Poem in the second Part of the Chym. Catalogue, but I conceive them to be in Prose; for you may see in Thor alias Thornly's Chiragogia [sic] Heliana, thirteen Ecchoes at the end of that Book," Sig. D* 24^v].

383 THRASTER [THRASHER], William
The Marrow of Chymical Physick; or the Practice of making Chymical Medicines; shewing the order to draw forth from Vegetables, Minerals, and Metals, their Spirits, Oyles, Vinegars, Salts, Extracts or Tinctures, Essences and Magisteries, Flowers, Salts, &c.
London, by T. J. for Peter Parker
1669
12°
Wing T1080; BL 1035. a. 17.

[In his Advertisement to the 1675 edition of the *Catalogue*, Cooper notes that in 1673 George Starkie's "Name was . . . put to the *Marrow of Chymical Physick*, instead of Will. Thrasher, who is the right author thereof."]

384 TONSTALL, George
Scarbrough Spaw Spagyrically Anatomized, together with a New-Years-guift (in verse) to Dr. Witty. (see WITTY, Robert, *Scarbroughs Spagyrical Anatomizer dissected*)
London, by J. M. for the Author
1671, 1672 [1670, Wing, BL]
8°
Wing T1889; BL 1171. a. 6. (2.)

[Wing and BL do not list the editions of 1671 and 1672 that Cooper reports, though Tonstall's *A New-year's Gift for Doctor Witty*, Wing T1888, is listed for 1672.]

Touch-stone for gold and silver wares (see BADCOCK, William)

Treasure of Treasures, or . . . the Golden Medicine (see HOUPREGHT, *Aurifontina Chymica*)

Treatise how Metals may be altered and transformed (see PORTA, *Natural Magick*, Books 5, 6, 7, and 10)

Treatise of the Philosophers Stone, the Blessed Manna (see HOUPREGHT, *Aurifontina Chymica*)

TREVISAN, Bernard, Earl of
Answer to the Epistle of Thomas of Bononia (see HOUPREGHT, *Aurifontina Chymica*)

TREVISAN, Bernard, Earl of
Treatise (see COOPER, *Collectanea Chymica*)

Dr. Trigs Water, with its vertues and use (see PARACELSUS, *Chymical Transmutation*)

385 *True and perfect order to distill oyles out of al maner of spices . . . gathered out of sundry aucthors* [sic]. Trans. J. H[ester]. (see PARACELSUS, *Key of Philosophy*)
London, J. Charlewood
1575
8°
STC 19181. 3

[Included in Paracelsus' *Key of Philosophy*, 1575+ editions, but is not by Paracelsus (STC).]

386 TRYE, Mrs. Mary
Medicatrix; or the Woman-Physician, vindicating T. O'Dowde, a chymical Physician . . . Chymistry and Chymical Medicines, and their Cures, and Promoting the Chymical Society in defence of Thomas O'Dowde against the calumnies of Henry Stubbe, &c.
London, by T. R. & N. T., and sold by Henry Broome, and John Leete
1675
8°
Wing T3174; BL 780. a. 40. (1.)

Tumba Semiramidis (see *COOPER, Collectanea Chymica*)

387 TURNER, William
**A new Boke of the natures and properties of all Wines that are commonlye used here in England . . . Whereunto is annexed the booke of the natures and vertues of Triacles, newly corrected and set foorth againe.*
London, W. Seres
1566 [1568, STC, BL]
8°
STC 24360; BL C. 31. c. 43.

388 TURNER, William
The rare treasure of the English Bathes: written by William Turner . . . Gathered and set forth for the benefite of his friendes and Countreymen in England by William Bremer. Pp. 110. [Included in Thomas Vicary, *The Englishemans Treasure: with the true Anatomie of mans bodie, etc.*]
London, imprinted by George Robinson for John Perin
1587
4°
STC 24708; BL C. 54. bb. 11.

[The folio edition of 1572 which Cooper lists is not included in the STC or the BL.]

389 TYMME, Thomas
A Dialogue Philosophicall. Wherein Natures secret Closet is opened, and the cause of all motion in Nature shewed out of Matter and Forme . . . Together with the wittie invention of an Artificiall perpetual motion.
London, printed by T. S., for Clement Knight
1612
4°
STC 24416; BL 8467. aa. 22.

390 VALENTINE, Basil
Basilius Valentinus . . . His Last Will and Testament. 5 pt.
London
1657
8°
Wing B1015; BL 8905. aa. 12.

391 VALENTINE, Basil
The Last Will and Testament of Basil Valentine . . . to which is added two treatises. The first declareth his manual operations. The second sheweth things natural and supernatural, etc. Pp. 534.

London, S. G. & B. G. for Edward Brewster
1670
8° [12°, Wing]
Wing B1017; BL C. 76. a. 22.

392 VALENTINE, Basil
Basilius Valentinus . . . Of Natural & Supernatural Things. Also, of the first Tincture, Root, and Spirit of Metals and Minerals, how the same are conceived, generated, brought forth, changed and augmented. Whereunto is added Roger Bacon, of the Medicine or Tincture of Antimony; Mr. John Isaac Holland, his Work of Saturn, and Alex. van Suchten, of the Secrets of Antimony. Translated by Daniel Cable. Pp. 238.
London, Moses Pitt
1671
8°
Wing B1020; BL 8905. a. 9.

VALENTINE, Basil
Salt of Antimony (see SUCHTEN, *Secrets of Antimony*)

393 VALENTINE, Basil
Triumphant Chariot of Antimony, being a conscientious discovery of the many real transcendent excellencies included in that Mineral, with a Discourse of what the Stone of Fire is . . . Faithfully Englished by J. H. Oxon. [i.e., John Harding]. Pp. 175.
1656: [no identifiable edition]
1660: Oxford, by A. Lichfield, for Thomas Bruster of London, 8° Wing B1021; BL 1606/102.
1661: London, printed for W. S., sold by Samuel Thomson, 8° Wing B1022; BL E. 2267. (2.)
1663: [no identifiable edition]
*1678: ["with the true book of the learned Synesius . . . concerning the Philosopher's Stone and annotations of Theodore Kirkringius, M. D."], London, for Dorman Newman, 8° Wing B1023; BL 1033. a. 4.

VAN HELMONT, Jean Baptiste (see HELMONT)

394 VAUGHAN, Thomas [PHILALETHES, Eugenius]
Anima Magica Abscondita; or a Discourse of the universall Spirit of Nature, With his strange, abstruse, miraculous Ascent, and descent. By Eugenius Philalethes.

London, by T. W. for H. B.
1650
8°
Wing V142; BL E. 1302. (2.)

395 VAUGHAN, Thomas [PHILALETHES, Eugenius]
Anthroposophia Theomagica; or A Discourse of the Nature of Man and his state after death; Grounded on his Creator's Proto-Chimistry, and verified by a practicall Examination of Principles in the Great World. By Eugenius Philalethes. 2pt.
London, by T. W. for H. Blunden
1650
8°
Wing V143; BL E. 1302. (2.)

[In the *Continuation*, Cooper also reports a 1656 edition of this work.]

396 VAUGHAN, Thomas [PHILALETHES, Eugenius]
Aula Lucis, or, the House of Light: A Discourse written in the year 1651. By S. N. A modern speculator.
London, for William Leake
1652
8°
Wing V144; BL 1033. d. 14.

397 VAUGHAN, Thomas [PHILALETHES, Eugenius]
A brief Natural History, intermixed with variety of philosophical discourses and observations of the Burnings of Mount Aetna. With refutations of such vulgar errours as our modern authors have omitted. By Eugenius Philalethes.
London, for Matthew Smelt
1669
8°
Wing V145; BL 970. e. 30. (1.)

398 VAUGHAN, Thomas [PHILALETHES, Eugenius]
Euphrates, or the Waters of the East; Being a short Discourse of that Secret Fountain whose Water flows from the Fire; and carries in it the Beams of the Sun and Moon. By Eugenius Philalethes.
London, for Humphrey Moseley
1655: 8° Wing V146; BL 445. a. 41.
1671: 8° Wing V147.

399 VAUGHAN, Thomas [PHILALETHES, Eugenius]
Fama Fraternitatis. The Fame and Confession of the Fraternity of R: C: commonly, of the Rosie Cross. With A Praeface annexed thereto, and a short Declaration of their Physical Work. By Eugenius Philalethes.
London, by J. M. for Giles Calvert
1652
8°
Wing V148; BL E. 1291. (3.)

400 VAUGHAN, Thomas [PHILALETHES, Eugenius]
A Hermeticall Banquet, drest by a Spagyricall Cook for the better preservation of the Microcosme.
London, for Andrew Crooke
1652
8°
Wing V149

401 VAUGHAN, Thomas [PHILALETHES, Eugenius]
Lumen de Lumine; or A new Magicall Light discovered, and Communicated to the World.
London, by H. Blunden
1651
8°
Wing V150; BL 234. a. 22. (3.)

402 VAUGHAN, Thomas [PHILALETHES, Eugenius]
Magia Adamica: or The Antiquitie of Magic, and the Descent thereof from Adam downwards, proved. Whereunto is added a perfect, and full Discoverie of the true Coelum Terrae, or the Magician's heavenly Chaos, and first Matter of all Things.
London
1650: by T. W. for H. Blunden, 8° Wing V151; BL E. 1299. (1.)
1656: J. Allen, 8° Wing V152; BL G. 19110. (1.)

403 VAUGHAN, Thomas [PHILALETHES, Eugenius]
The Man-Mouse taken in a Trap, and tortur'd to death for gnawing the margins of Eugenius Philalethes. Pp. 116.
London, for H. Blunden
1650
8°
Wing V153; BL E. 1299. (2.)

404 VAUGHAN, Thomas [**PHILALETHES, Eugenius**]
Second Wash; or the Moore Scour'd once more; being a Charitable Cure for the Distractions of Alazonomastix.
London, by T. W.
1651
8°
Wing V154; BL 526. f. 14. (5.)

405 VENNER, Tobias
The Baths of Bathe; or, a necessary compendious Treatise concerning the nature, use, and efficacie of those famous hot waters; with a Censure concerning the water of St. Vincents Rock neere Bristoll, and its use against the Stone.
London, F. Kyngston for R. Moore
1628
4°
STC 24641; BL 1171. d. 46.

[Cooper also reports inclusion of this title in Venner's *Via Recta ad vitam longam*, fourth ed., London, for Abel Roper, 1660, 4° Wing V196.]

406 **A very proper treatise, wherein is briefly sett forthe the arte of Limming, etc.* [Later editions of 1573 work.]
London, T. Purfoot
1596: 4° STC 24256; BL C. 31. e. 37.
1604 [1605, STC, BL]: 4° STC 24257; BL C. 31. c. 25.

407 VIGENERE, Blaise de
A discourse of Fire and Salt, discovering many secret mysteries, as well philosophicall as theologicall [translated by Edward Stephens]. Pp. 162.
London, A. Crooke [Richard Cotes, Wing]
1649
4°
Wing B3128; BL 1033. g. 38.

VILLINGANUS, George Pictorius
Discourse of the Nature of Spirits (see AGRIPPA, *Fourth Book*)

408 W[ALWIN], **W**[illiam]
**Health's New Store-house opened, offering to familiar use such supplies as are most wanting . . . to humane frailty.*
London, by Jane Clowes, to be sold by John Sweeting
1661

8° [4°, Wing, BL]
Wing W147; BL 7391. de. 48.

[Cooper attributes this work to William Walwyn.]

409 WALWYN, William
**Physick for Families* [Wing and Cooper say "familiar"]: *or, the new, safe and powerful way of physick, upon constant proof established*, etc. Pp. 144.
London
1669: by J. Winter, to be sold by Robert Horn, 8° Wing W687; T. C. I 25.
1681: by J. R., to be sold by John Starky, 12° [8°, BL] Wing W689; BL C. 54. a. 26.

410 WARD, Samuel
The Wonders of the Loadstone, or the Load-Stone newly reduc't into a divine and morall use . . . [Translated from the Latin by Sir H. Grimston].
London, E. Purslow for P. Cole
1640
12°
STC 25030; BL 8756. a. 27.

411 WARREN, Hardick
**Magic and Astrology vindicated from false aspersions and calumnies.*
London, J. M. for N. Brook
1651
4°
Wing W971; BL E. 623. (4.)

Water-stone of the Wise Men (see SIEBMACHER, Johann)

412 WEBSTER, John
Metallographia, or an history of metals, with Signs of Ores, and minerals, before and after diging, the causes and manner, of their Generations, their kinds, sorts, and differences, with a description of new metals, and Semi metals, and other things pertaining to mineral knowledge, also of Vegitability, of mystical Chymistry, of the Philosophers Gold and Mercury, of the Liquor Alkahest, Aurum Potabile, and such like.
London, A. C. for Walter Kettilby
1671
4°

Wing W1231; BL 990. e. 1.

413 WECKER, Hans Jacob
Eighteen books of the Secrets of Art and Nature, being the sum and substance of Natural Philosophy, methodically digested, much augmented and enlarged by Dr. R. Read, etc.
London, for Simon Miller
1660
fol.
Wing W1236; BL 64. e. 11.

414 WEIDENFIELD, Johan Seger von
**Four Books . . . concerning the Secrets of the Adepts; or, of the use of Lully's spirit of wine. A practical work, . . . collected out of the ancient as well as modern fathers of adept philosophy.* [Translated from the Latin by G. C.]
London, by Will. Bonny, for Thos. Howkins
1685
4°
Wing W1253; BL 1033. i. 18. (1.)

415 WHITAKER, Tobias
*Περι υδροποσιας : *or, a discourse of waters: their qualities and effects.*
London, printed for J. Grismond
1634
12°
STC 25355; BL 1171. a. 5.

416 WHITAKER, Tobias
**The Tree of humane life, or the bloud of the Grape, proving the possibilitie of maintaining human life from Infancy to extreame old age without any sicknesse by the use of wine.*
1638: London, printed by J. D[awson] for H. O[verton], 8° STC 25356; BL 1038. b. 17.
1654: [second edition], London [Hague, Wing], for Samuel Broun, 12° Wing W1714

417 WHITE, John
**Art's Treasury; or, a profitable . . . invitation to the Lovers of Ingenuity; contained in many extraordinary experiments, rarities and curious inventions.*
London, for W. Whitwood; and Mrs. Feltham

1688
12°
Wing W1788; BL 56. a. 30.

418 WILKINS, John
Mathematicall Magick; or, the wonders that may be performed by mechanicall geometrie. In two books.
London, by M. F. for S. Gellibrand
1648
8°
Wing W2198; BL E. 1095.

419 WILLIAMS, Ralph
Physical Rarities: containing the most choice receipts of physick and chyrurgerie, for the cure of all diseases stored with admirable Secrets.
London
1651: for J. M. to be sold by George Calvert, 8° Wing W2751; BL E. 1302. (1.)
1652: [second edition to which is added *Hermes Trismegistus his Physical Mathematicks*], for W. L. and J. M. to be sold by W. Ley and George Calvert, 8° Wing W2752
*1657: [third edition], for William Ley, 8° Wing W2753; BL 1038. d. 38. (1.)

420 WILLIAMS, William
Occult Physick; or three principles of Nature Anatomized, by a Philosophical operation, from Experience in three Books of Beasts, Trees, Herbs, and their Magical and Physical vertues.
London, by Tho. Leach, to be sold by W. Palmer
1660
8°
Wing W2784; BL E. 1737. (2).

421 WILLIS, Timothy
The Search of Causes. Containing a Theophysicall investigation of the possibilitie of transmutatorie Alchemie.
London, J. Leggatt
1616
8°
STC 25754; BL 1036. a. 13. (2.)

422 WILLSFORD, Thomas

**Nature's secrets; or, the admirable and wonderfull history of the generation of meteors; particularly describing the temperatures and qualities of the four elements.*
London, for Nath. Brook
1665 [1658, Wing, BL]
8°
Wing W2875; BL E. 1775. (2.)

423 WILSON, Edward
**Spadacrene Dunelmensis: or a short treatise of an ancient medicinal fountain or vitrioline Spaw near . . . Durham.*
London, by W. Godbid
1675
8°
Wing W2891; BL 1171. a. 27.

424 WILSON, George
**Sal est bonum Dei donum, &c. or the admirable, universal, generative, preservative and restorative Balsom, Chymically prepared, or the odoriferous Spirit of Salt, its Virtues and Uses, also the true Oyl or Spirit of Sulphur, per Campanum, and its Virtues in Physick, with a Catalogue of Mr. Wilson's Chymical Medicines.* [2 sheets]
London
1665
8°

[The *Catalogue* is an earlier issue of the broadside, *Gaza Chymica: or, A Magazin, or Store-house of Choice Chymical Medicines: Faithfully prepared . . . by George Wilson, Philo-Chym.*, London, 1686, Wing W2894; BL 777. k. 16. (15.)]

425 WINE, Richard
**The poor man's mite, cast into the Treasury of Endeavours.*
London, by Tho. Newcomb
1660
4°
Wing W2982

426 WITTY, Robert
Pyrologia Mimica; or, an answer to Hydrologia Chymica of W. Sympson; in defence of Scarbrough Spaw, etc.
London, by T. N. for J. Martyn
1669

8°
Wing W3230; BL 233. a. 52. (2.)

427 WITTY, Robert
**Scarbroughs Spagyrical Anatomizer dissected. Or an answer to all that Dr. Tonstall hath objected in his book against Scarborough Spaw . . . As also reflections upon a late piece [by W. Simpson] called a Vindication of hydrologia chymica.* (see TONSTALL, George and SIMPSON, William)
London, by B. G. for Nath. Brooke and R. Lambert in York
1672
8°
Wing W3233; BL 7441. a. 42.

428 WITTY, Robert
Scarbrough Spaw; or, a description of the nature and vertues of the Spaw at Scarbrough in Yorkshire. Also a treatise of the nature and use, of Sea, Rain, Snow, Pond, lake, Spring and River-waters, with a discourse concerning Mineral-water, and the several arts thereof, etc.
1660: London, Charles Tyus and Richard Lambert, 8° Wing W3231; BL E. 1830. (2.)
*1667: [second edition, "with the Addition of an Historical Relation of the Cures done by those Waters"], York, by A. Broad for Tho. Passenger: to be sold by Richard Lambert, 8° Wing W3232; BL 1171. a. 10.

ZOROASTER
Cave (see THOR, *Cheiragogia Heliana*)

ZOROASTER
Chaldaick Oracles (see STANLEY, Thomas)

Appendix I

Supplementary Entries from Lists Included in Other Works Published by William Cooper

In addition to the titles included in the two versions of the *Catalogue of Chymicall Books* (1673, 1675) and its *Continuation or Appendix* (1688), it was Cooper's practice to print, in several of the volumes that he published, supplementary lists of books available for purchase at his bookshop. Thus he adopted an advertising method that was common among publishers of his time as well as our own. Bearing such headings as "Books Sold by William Cooper" and "A Catalogue of Certain Chimical-Books Sold by William Cooper at the Pelican in Little-Britain, London," these lists are important evidence of his continuing involvement in the book trade between 1675 and 1688, years which, as we have seen, coincide exactly with his book auctioneering.

Given the distinct possibility that Cooper's decision to publish the *Continuation or Appendix* may have been something of an afterthought (see Introduction), it is likely that, besides their primary value as advertising, he also regarded these supplementary lists as means of updating the *Catalogue*. Through them readers could be apprised of newer books currently available for purchase at the Pelican; in addition, they might serve as a ready basis for a more complete and formal compilation in the event that one was desired. The great majority of the titles that first appeared in these lists, did, in fact, show up again in the *Continuation*.

Whatever Cooper's motives, his emphasis is on works that appeared following publication of the expanded two-part *Catalogue* of 1675. For example, of the total of 105 separate titles contained in these lists (some are present on more that one), 41 are works published after 1674. Among the authors and titles that appear for the first time, few can be said to represent subjects that are positively new; Thomas Ash's *Carolina* (Wing A3934), John Marbecke's *Lives of the holy Saints* (Wing M565F), John Owen's *Truth and Honesty in Plain English* (Wing T3149), and *Spencer's* [sic] *View of Ireland*, probably in a folio edition of 1633 (STC 25067), are clearly departures from Cooper's more abiding interests. The latter are well represented by new works by Robert Boyle as well as those that passed confusingly under the names of Philalethes and George Starkey (see also Appendix II).

Among these book lists, the most complete and informative are found in three important collections of alchemical treatises which he either compiled and edited or had printed for sale at the Pelican. *Ripley Reviv'd* (1678; Wing S5286), by Eirenaeus Philalethes, comprises six analytical treatises on the "Hermetico-Poetical" writings of the fifteenth-century English alchemist, George Ripley. The *Aurifontina Chymica* of 1680 (Wing H2941), contains fourteen treatises, as well as an epistle dedicatory by John Frederick Houpreght, who also translated its first essay, the *Hydropyrographum Hermeticum*. Finally, the *Collectanea Chemica: A Collection of Ten Several Treatises in Chymistry* (1684; Wing C5103), includes an epistle to the reader which is signed W. C. B., i.e., William Cooper Bookseller.

The entries included in this appendix represent the entire contents of four separate lists scattered throughout the *Collectanea Chemica* and one each from the *Aurifontina Chymica* and *Ripley Reviv'd*. Specific locations within these collections are identifiable by means of the following key:

1 Printed following the *Speculum Alchymiae; The True Glass of Alchemy*, written by Roger Bacon and printed in the *Collectanea Chemica*, pp. 134–6.

2 Printed following "The Admirable Efficacy, and almost incredible virtue of true Oyl" in the *Collectanea Chemica*, p. 152.

3 Printed following "A Petition to the courteous Reader" in the *Collectanea Chemica*, pp. 196–97.

4 Printed following "A Treatise of Bernard, Earl of Trevisan, Of the Philosopher's Stone," in *Collectanea Chemica*, pp. 95–8.

5 Printed at the end of the *Aurifontina Chymica*, p. 272.

6 Printed as the three final pages of *Ripley Reviv'd*.

ANON. (6)
Lust's Dominion; or, the lascivious Queen, a tragedie, etc. [In five acts, in verse, and prose. Edited by F. Kirkman.] Pp. 150.
London, for F[rancis] Kirkman and are to be sold by Robert Pollard [Doubtful and supposititious attribution to Christopher Marlowe.]
1657
12°
Wing L3504A; BL 643. a. 28.

ANON. (1)
The Practice of Lights: or, an Excellent and ancient treatise of the philosopher's stone (see COOPER, *Collectanea Chemica*)

APOLLONIUS, Tyaneus (2)
Ars Notoria: the Notory Art of Solomon, shewing the cabalistical key of magical operations, the liberal sciences, divine revelation, and the art of memory. Whereunto is added an Astrological Catechism, fully demonstrating the art of Judicial Astrology . . . Written originally in Latine [by Apollonius, Leovitius, and others. Collected] *and now Englished by R. Turner.*
London, by J. Cottrel, to be sold by Martha Harison
1657
8°
Wing A3533; BL E. 1655. (2.)

(5)
The Art of Metals, how to find, know and refine them from their Ore.

[Additional information unavailable.]

ASH, Thomas (2)
Carolina: or a description of the present state of that country . . . by T. A., Gent. [i.e., Thomas Ash?]
London, for W. C., to be sold by Mrs. Grover
1683 [1682, BL and Wing]
4°
Wing A3934; BL 1061. g. 60.

ASHMOLE, Elias (3)
The Way to Bliss. In three books, etc. Pp. 220.
London, John Grismond for Nath. Brooke
1658
4°

Wing A3988; BL E. 940. (3.)

BACON, Roger (4)
The Cure of Old Age, and Preservation of Youth . . . Translated out of Latine [i.e., from the abridgement published in 1590]. . . *Also a Physical account of the tree of life by Edw. Madeira Arrais, translated likewise out of Latin.* 2 pt.
London, for Tho. Fletcher and Ed. Evets
1683
8° [12°, BL]
Wing B372; BL 1039. d. 22.

BACON, William (4)
A Key to Helmont. Or, a short introduction to the better understanding of the theory and method of the most profound chymical physicians. Pp. 34.
London, for John Starkey
1682
4°
Wing B374; BL 1034. h. 35.

BADCOCK, William (4)
A New Touchstone for Gold and Silver Wares [second edition].
London, for John Bellinger and T. Bassett
1679
8°
Wing B381; BL 1651/1059.

BADCOCK, William (4)
A Touch-stone for Gold and Silver Wares. Or, a manual for goldsmiths . . . by W. B.
London, for John Bellinger & Thomas Bassett
1677
8°
Wing B382; BL 1651/170.

BARBA, Alvaro Alonso (1)
Two Books of the Art of Mettals; In the first is declared the manner of their Generation, and the concomitants of them. In the Second is taught the common way of refining Silver by Quick-silver; with some new Rules added for the better performance of the same: translated out of the Spanish into English by Edward Earl of Sandwich. 2 vols.

London, for S. Mearne
1674
8°
Wing B680/682; BL 446. a. 25. (3).

BECHER, Johann Joachim (4)
Magnalia Naturae: or, the Philosophers-stone lately expos'd to public sight and sale. Being a true and exact account of the manner how Wenceslaus Seilerus . . . came by and made away with a very great quantity of powder of projection. Pp. 31.
London, by Tho. Dawks. Sold also by La. Curtiss
1681 [1680, BL and Wing]
4°
Wing B1643; BL 1035. c. 42.

BERNHARD, Earl of Trevisan (4)
Epistle to Thomas of Bononia, concerning the secret workings of Nature in the product of things (see HOUPREGHT, *Aurifontina Chymica*)

BOEHME, Jacob (1,5,6)
Aurora. That is, the Day-spring, or dawning of the day in the orient, or the morning redness in the rising of the Sun. That is the root or mother of philosophie, astrologie, and theologie, from the true ground; Or a description of Nature as she was originally, how changed creaturally, how she stands at present, and will be at the end, [Translated by J. Sparrow]. Pp. 643.
London, John Streater for Giles Calvert
1656
4°
Wing B3397; BL E. 890.

BOEHME, Jacob (5)
The Forty Questions of the Soul.
London, for L. Lloyd
1665
8°
Wing B3407;BL 3716. aa. 27.

BOEHME, Jacob (1,4,5)
The Remainder of Books written by J. Behme, viz. I. The First Apologie to Balthazar Tylcken . . . II. The Second Apologie in answer to Balthazar Tylcken . . . III. The Fouer Complexions . . . IV. The Considera-

tions upon Esaiah Stiefel's Booke concerning the Threefold State of Man and the New Birth . . . V. The Apologie in answer to Esaiah Stiefel concerning perfection . . . VI. The Apologie in answer to Gregory Rickter . . . for the way to Christ . . . VII. Twenty five epistles more then the 35 formerly printed in English, with 2 . . . which make 62 in all, also 1 epistle more of his own hand writing; and 1 of Dr. Charles Weisners [i.e., Cornelius Weissner], *relating much of J. B.'s life. Englished by John Sparrow* (with a bibliography). 7 pt.
London, by M. S. for Giles Calvert
1662
4°
Wing B 3415; BL 853. h. 13.

BOYLE, Robert (4)
Aerial Noctiluca: or, some new phoenomena, . . . the wonder of this Age. Pp. 109.
London, by T. Snowden, and are to be sold by Nath. Ranew
1680
8°
Wing B3925; BL 1034. g. 30.

BOYLE, Robert (1,5)
Essay about the Origine & Virtues of Gems. Wherein are propos'd and historically illustrated some conjectures about the consistence of the matter of precious stones, and the subjects wherein their chiefest virtues reside, Pp. 185.
London, printed by William Godbid; sold by Moses Pitt
1672
8°
Wing B3947; BL 990. b. 2.

BOYLE, Robert (1,5)
Essays of the Strange Subtilty, Determinate Nature, Great Efficacy of Effluviums. To which are annext New experiments to make fire and flame ponderable: together with a Discovery of the perviousness of glass. 4 pt.
London, Wm. Godbid for M. Pitt
1673
8°
Wing B 3951; BL 535. b. 19.

BOYLE, Robert (4)

[*An Historical Account*] *of a Degradation of Gold, made by an anti-elixir: a strange chymical narative* [sic]. Pp. 17.
London, by T. N., for Henry Herringham
1678
4°
Wing B3984; BL 1035. c. 44.

BOYLE, Robert (4)
New Experiments and Observations made upon the Icy Noctiluca. Pp. 150.
London, by R. E. for B. Tooke
1682 [1681/82, Wing]
8°
Wing B3995; BL 1034. g. 31.

BOYLE, Robert (1,5)
Observations about the Growth of Metals in their Ore exposed to the Air. (see Tracts: containing 1. Suspicions about some hidden qualities of the air, etc. [in pt. 2 of 6 pts.])
London, W. G. [William Godbid]; sold by M. Pitt
1674
8°
Wing B4054; BL 535. b. 21.

BRATHWAITE, Richard (6)
Tragi-Comoedia, cui in titulum inscribitur Regicidium, perspicacissmis judiciis acuratius perspecta, pensata, comprobata. (Bedlamum novum. Scena Britannia.) 2pt. Pp. 190.
Londini, typis J. G[rismond] & prostat venalis in officina Theodori Sadleri
1665
8°
Wing B4278; BL C. 77. a. 19.

BURGGRAV, Johannes E. (4,5,6)
The Method of chymical philosophie and physick. Being a brief Introduction to the one, and a true discovery of the other.
London, by J. G. for Nath. Brook
1664
8°
Wing M1943; BL 1651/1227

[Cooper refers to this work as *Introduction to Astral Philosophy.*]

CASE, John (4)
Wards of the Key to Helmont proved Unfit for the Lock; or, the Principles of Mr. W. Bacon examined and refuted, and the honour and value of true chymistry asserted. Pp. 24.
London, for the Author, sold by John Smith
1682
4°
Wing C821; BL 1034. h. 36.

CHARAS, Moise (1)
Royal Pharmacopoea, Galenical and chymical, according to the practice of the most eminent and learned physitians of France, etc. Pp. 272. 245.
London, for John Starkey and Moses Pitt
1678
fol.
Wing C2040; BL 547. l. 11.

COLLINS, John (3)
Salt and Fishery, a discourse thereof, etc. Pp. 164.
London, by A. Godbid and J. Playford; sold by Robert Horne.
1682
4°
Wing C5380; BL 981. b. 5.

COOPER, William (4)
A Catalogue of Chymicall Books. In Three parts. In the First and Second Parts are contained such Chymical Books as have been written Originally, or Translated into English: With a large Account of their Titles, several Editions and Volumes. Likewise In the Third Part is contained a Collection of such things published in the Philosophical Transactions of the Royal Society (for Ten Years Together) as pertain to Chymistry, or the Study of Nature by Art in the Animal, Vegetal, and Mineral Kingdoms. [For an earlier edition, see COOPER, *Philosophical Epitaph.*]
London, T. R. & N. T. for William Cooper
1675
8°
Wing C6061; BL G. 19086.

C[OOPER], W[illiam] (1)
Collectanea Chymica: a collection of Ten Several Treatises in Chymistry, concerning the Liquor Alkahest, the Mercury of Philosophers, and other Curiosities worthy the Perusal. Written by Eir. Philaletha, Anonymus, Joh. Bapt. Van-Helmont, Dr. Fr. Antonie, Bernhard Earl of Trevisan,

Sir Geo. Ripley, Rog. Bacon, Geo. Starkey, Sir Hugh Platt, and the Tomb of Semiramis. [The epistle to the reader signed: W. C. B., i.e., William Cooper, Bookseller.]
London, Printed for William Cooper, at the Pelican in Little Britain
1684
8°
Wing C5103; BL 1033. d. 15.

["Notwithstanding the Latin titles the tracts are all in English, except the first, which is in both languages. Each tract has a title page," Ferguson, I, 169.]

Contents:
1. Eirenaeus PHILALETHA, *Arcanum, or Secret of the Immortall Liquor Alkahest*, p. 5.
2. ANONYMOUS, *Practice of Lights*, p. 27.
3. Joh. Bapt. Van HELMONT, *Praecipiolum*, p. 47.
4. Fr. ANTONIE, *Aurum Potabile*, p. 73.
5. BERNHARD of Trevisan, *Treatise*, p. 83.
6. George RIPLEY, *Bosome-Book*, p. 101.
7. Roger BACON, *Speculum Alchymiae*, p. 125.
8. George STARKEY, *Oil of Sulphur-vive*, p. 137.
9. Hugh PLATT, *Remedies against Famine*, p. 153.
10. H. V. D., *Tumba Semiramidis*.

COOPER, William (1,5,6)
The Philosophical Epitaph of W. C. Esquire. For a Memento Mori on his Tomb-stone.
London, T. R. & N. T. for William Cooper
1673
8°
Wing C6062; BL 1033. g. 22. (1.)

Contents:
1. *Three Hieroglyphical Scutcheons displaying Minerva's and Hermes Birds, and Apollo's Bird of Paradise in Philosophical Mottos and Sentences with their Explication, and a discovery of the Liquor Alchahest, of Salt of Tartar volatized, and other Elixirs, with their differences.*
2. *A Briefe of the Golden Calfe, discovering the rarest Miracle in Nature, of a transmutation of Lead into Gold made by Dr. Jo. Fred. Helvetius with Figures.*
3. *Jo. Rod. Glauber his Golden Ass well managed, and Midas restored to Reason, a new Chymical light for comfort of the oppressed, demon-*

strating Gold to be easily extracted in all places out of Sand, Stones, Gravel, or Flints, and the true matter of the Philosophers Stone.
4. *Jehior, the day Dawning, or the Morning Light of Wisdom, containing the three Principles or Originals of all things whatsoever, discovering the great and many Mysteries in God, Nature, and the Elements. All published by W. C., Esq. With a Catalogue of Chymical Books (collected by Will. Cooper, Bookseller).*

COWELL, John (6)
The Interpreter: or Booke containing the Signification of Words: wherein is set foorth the true meaning of all, or the most part of such words and termes, as are mentioned in the lawe writers, or statutes of this victorious and renowned kingdome, requiring any exposition or interpretation, etc.
Cambridge, John Legate
1607
4°
STC 5900; BL 507. d. 14.

COX, Francis (3)
Treatise of the making and use of divers Oyls, distilled Waters, &c.
London
1575
8°

[Additional information unavailable.]

CROLLIUS, Oswald (1,5)
Admonitory Preface of Oswald Crollie, Physitian: to the most illustrious Prince Christian Anhaltin. [A Translation of Crollius, *Basilica Chymica*, by H. Pinnell under the title, *Philosophy reformed and improved in four profound tractates*]. 4pt.
London, by M. S. for Lodowick Lloyd
1657
8°
Wing C7023; BL E. 1589. (1.)

CROLLIUS, Oswald (1)
Bazilica chymica, & Praxis chymiatricae. Or, Royal and practical Chymistry, in three treatises . . . Being a translation of Oswald Crollius his Royal Chymistry, augmented and inlarged by John Hartman. To which is added his Treatise of Signatures of Internal Things . . . As also The Practice of Chymistry of John Hartman M. D. augmented and inlarged

by his son. All faithfully Englished by a lover of chymistry. 3 pt.
London, for John Starkey, and Thomas Passinger
1670
fol.
Wing C7022; BL C. 113. h. 8

DIGBY, Sir Kenelm (4)
A Choice Collection of Rare Secrets and Experiments in Philosophy. Collected and experimented by . . . Sir Kenelm Digby . . . now . . . published . . . by Mr. Hartman his Steward and Operator.
London, to be sold by W. Cooper, Henry Farthorne, and John Kersey
1682 [and reissued]
8°
Wing D1425A, D1426, D1426A; BL C. 112. aa. 14

E., Y. (5,6)
A Caveat for the Protestant Clergy, if Popery be restored.
n.p.
1671
8°
Wing E45; Bodleian

(2)
Eirenaeus Philalethes Exposition in Gebrum, de Praeparatione Mercurii Sophici & breve Manuductorium ad Campum Sophiae
(see PHILALETHES, AEyrenaeus, *Opus Tripartitum*)

FABRICIUS, Joannes Ludovicus[FERRARIUS, Janus Alexander, pseud.] (6)
Jani Alexandri Ferrarii Euclides Catholicus, sive demonstratio Romanae fidei ex primis, certi et evidentibus principiis, mathematica methodo et connexis continua serie propositionibus deducta. Pp. 63 [i.e., 86]
London, Prostant venales apud Guliel. Gilbert
1673
4°
Wing F74; BL 475. a. 4.

FELGENHAUER, Paul
Jehior, the day Dawning, or the Morning Light of Wisdom (see COOPER, William, *Philosophical Epitaph*)

FESTEAU, Paul (1,6)

A French Grammar.
London, by W. G. for William Cooper
1674
8°
Wing F823C; BL 1483. aa. 28.

(4)
Nicolas FLAMMELLs Summary of the Philosophers-Stone
(see HOUPREGHT, *Aurifontina Chymica*)

FLETCHER, Richard (3)
Character of a true Physician; or a true chymist compared with a goose quill pedant, etc.
London, for the Author
1676
8°
Wing F1357; BL 778. c. 39.

FLETCHER, Richard (3)
Starkey revived; or, collections out of nature's explication and Helmont's vindication.
London, for the author
1676
8°
Wing F1359; BL 780. b. 3. (4.)

FRANCISCUS, a Sancta Clara [DAVENPORT, Christopher] (6)
Deus, natura, gratia. Sive Tractatus de praedestinatione, de meritis & peccatorum remissione, seu de iustificatione, & denique de sanctorum invocatione. Ubi ad trutinam fidei catholicae examinatur confessio anglicana, etc. Pp. 372.
Lugduni, sumptibus Antonii Chard
1634
4°
BL 4378. l. 12.

FROMMENIUS, Andreas [Andreas FROMMAN](6)
Synopsis Metaphysica Andreae Frommanii. . . .
Oxford, typis H. Hall, impensis Tho. Gilbert
1669
12° [8°, Wing]
Wing F2243; Bodleian

FULKE, William (4)
A Goodly Gallerye with a most pleasaunt Prospect, into the garden of naturall Contemplation, to behold the Naturall causes of all kynde of Meteors, as wel fyery and ayery, as watry and earthly, etc.
London, W. Griffith
1563
8°
STC 11435; BL C. 54. a. 25.

GLASER, Christophe (4)
Compleat Chymist, or a New Treatise of Chymistry . . . from the fourth edition revised and augmented by the Author. Now faithfully englished by a Fellow of the Royal Society.
London, for John Starkey
1677
8°
Wing G843; BL 1142. a. 36.

GLAUBER, John Rudolph (1)
Description of new Philosophical Furnaces, or, a new art of distilling, . . . Whereunto is added a description of the Tincture of Gold, or the true Aurum Potabile, [with] the first part of the Mineral Work.
London, by Ric. Cotes for T. Williams
1651
4°
Wing G846; BL E. 649. (3.)

GLAUBER, John Rudolph
Golden Ass well managed (see COOPER, *Philosophical Epitaph*)

GODFREY, Robert (2,5)
Various injuries & abuses in chymical and galenical physick committed both by Physicians and Apothecaries, etc.
London, by John Darby, for Richard Jones
1674
8°
Wing G927; BL 1038. a. 17.

HEAD, Richard (6)
The English Rogue described in the Life of Meriton Latroon, a witty extravagant. Being a compleat history of the most eminent cheats of both sexes, etc. 4 pt.

London, for Francis Kirkman
1665–68–71–72
8°
Wing H1248bA; BL 12614. c. 21.

[Cooper does not cite specific date for his edition.]

HELMONT, Franciscus Mercurius van (3)
Cabbalistical Dialogue concerning the Worlds Creation, in answer to . . . a learned Doctor [i.e., Henry More], etc.
London, for Benj: Clark
1682
4°
Wing H1390; BL 701. g. 47.

HELMONT, Jean Baptiste van (1)
Praecipiolum; or the immature-mineral electrum. The first metall: which is the mineral of Mercury. [For another edition, see COOPER, *Collectanea Chymica.*]
London
1683
8°
Wing H1400A

HELMONT, Jean Baptiste van
Van Helmont's Workes, containing his . . . Philosophy, Physick, Chirurgery, Anatomy; wherein the philosophy of the schools is examined, the errours therein refuted, and the whole art reformed . . . , done into English by J. C. [i.e., J. Chandler].
London, for Lodowick Lloyd
1664
fol.
Wing H1397; BL 549. l. 27.

HELVETIUS, Johann Friedrick (1)
Briefe of the Golden Calfe (see COOPER, William, *Philosophical Epitaph*)

HESTEAU, Clovis, Sieur de Nuysement
Sal, lumen, et spiritus mundi philosophici: or, the Dawning of the day, discovered by the beams of light: shewing the true salt and secret of the philosophers, the first and universal spirit of the world. Written originally in French [by Baron J. de Nuisement], *afterwards turned into Latin*

. . . by Lodovicus Combachius . . . and now transplanted into Albyons garden, by R. T. [Robert Turner]. Pp. 220.
London, by J. C. for Martha Harrison
1657
8°
Wing N1469; BL C. 145. f. 20.

HOUGHTON, Thomas (4)
Rara Avis in Terris; or, the Compleat Miner . . . the Liberties, Laws, and Customs of the Leadmines . . . in Derbyshire.
London, for William Cooper
1681
12°
Wing H2934; BL 445. a. 25. (1.)

HOUPREGHT, John Frederick (4,5)
Aurifontina Chymica: or, a Collection of Fourteen small Treatises Concerning the First Matter of Philosophers, for the discovery of their (hitherto so much concealed) *Mercury.*
London, for William Cooper
1680
24° [12°, BL]
Wing H2941; BL 1036. a. 29.[lost]; Wellcome 39689.

Contents:

1. *Hydropyrographum Hermeticum, or the Metallick Water-Fire, translated* by J. F. Houpreght, p. 1.
2. *The Privy Seal of Secrets, plainly discovering the first Matter of the Philosophers*, p. 41.
3. *A strange Letter concerning an Adept, his curious Learning and vaster Treasure, from Frederick, Duke of Holsatia and Sleswick*, p. 53.
4. Sir George Ripley's *Treatise of Mercury, and the Philosophers stone*, p. 69.
5. *Colours to be observed in the operation of the Great Work of Philosophers*, p. 93.
6. *A plain and true Description of the Treasure of Treasures, or the Golden Medicine*, p. 97.
7. *A Treatise of the Philosophers Stone, the Blessed Manna, &c., with its wondrous Virtues and Use, both for the curing of the Body of Man, and the making of the precious Stones*, p. 107.
8. Nic. Flammell, his *Summary of Philosophy, or Short Treatise of the Philosophers Stone*, p. 145.
9. Raym. Lullie's [Lull] *Clavicula, Apertorium, or little Key explaining*

all the rest of his Works, p. 163.
10. *Secrets disclos'd of the Philosophers Stone*, p. 180.
11. *A Philosophical Riddle of Gold, in Verse*, p. 185.
12. Bern. Trevisan's *Epistle to Thomas of Bononia, concerning the secret workings of Nature in the product of things*, p. 187.
13. Bern. Trevisan, his *short Epistle Dedicatory before his Works, to the same man*, p. 269.
14. *A brief Preparation of the Philosophers Stone, for the conclusion of this Book*, p. 271.

(4)
Hydropyrographum Hermeticum
(see HOUPREGHT, *Aurifontina Chymica*)

JABIR IBN HAIYAN (4,5)
The Works (or rather the Summa Perfectionis, Liber investigationis magisterii, De inventione veritatis and Liber Fornacum) of Geber . . . Englished by R. Russel.
London, for N. E. by Thomas James; to be sold by Robert Clavel
1678
8°
Wing J54; BL 236. i. 19.

JOHNSON, William (6)
Lexicon Chymicum, cum obscuriorum verborum et rerum Hermeticarum, tum phrasium Paracelsicarum in scriptus ejus, et aliorum chymicorum, passim occurrentium . . . explicationem contineus. (Vita Paracelsi.). 2 pt.
London, Excudebat G. D. impensis Gulielmi Nealand
1657
8°
Wing J856; BL 8906. de. 8.

KENDALL, George (3, 4)
An Appendix to the Unlearned Alchimist [of Richard Mathews], *wherein is contained the true Receipt* [for the preparation and making of] *Matthew's Pill*, (see MATHEWS, Richard, *Unlearned Alchymist*).
London, for Joseph Leigh to be sold by the author
1663 [1664, Wing, BL]
8°
Wing K283; BL 1033. d. 2. (2.)

KHRYPFFS, Nicolaus (4)
The Idiot in four books; the first and second of wisdome, the third of the minde, the fourth of statick experiments, etc. By the famous and learned C. Cusanus. (Translated from the Latin.)
London, for Wm. Leake
1650
12°
Wing K394; BL 1248. a. 34.

LEMERY, Nicolas (4)
An Appendix to a Course of Chymistry . . . Together with the process of the Volatile Salt of Tartar, and some other useful preparations. Writ in French by M. N. Lemery. Translated by W. Harris, M. D.
London, for Walter Kettilby
1680
8°
Wing L1037A; BL 1651/1036. (2.)

LEMERY, Nicolas (4)
A Course of Chymistry; containing the easiest manner of performing those Operations that are in use in Physick . . . Writ in French by M. N. Lemery. Translated by Walter Harris, M. D.
London, for Walter Kettilby
1677
8°
Wing L1038; BL 1651/1036. (1.)

LULL, Ramon (4)
Clavicula, or Key to his Works
(see HOUPREGHT, *Aurifontina Chymica*)

MACKAILE, Matthew (3)
The Diversitie of Salts and Spirits maintained. Or, the imaginary volatility of some salts and non-entity of the alcali, before cremation, and identity of the alcalies, all volatil salts, and all vinous spirits, by an onely lamp-furnace, resolved into real improbability. By way of animadversions upon Dr. D. C[oxe] *his 3 papers, communicated to the R*[oyal] *S*[ociety] *and insert in the 9 vol. of the P*[hilosophical] *T*[ransactions]. *Together with a new system of the order and gradation in the world's creation. As also Scurvie Alchymie discovered.* (see COXE, Daniel, *Discourse.*)
Aberdeen, by John Forbes
1683
8° [12°, BL]

Wing M144; BL 1142. a. 35.

MACKAILE, Matthew (3)
Moffet-Well; or, a topographical-spagyricall description of the mineral wells at Moffet in Annedale in Scotland . . . as also the Oyly-Well at St. Catherines Chappel in the Paroch of Libberton.
Edinburgh, for Robert Brown
1664
8°
Wing M148; BL 1171. e. 43.

MADEIRA ARRAIS, Edward (4)
A Physical account of the tree of life (see BACON, Roger, *The Cure of Old Age, and Preservation of Youth*)

MAGNENUS, Joannes Chrysostomus (6)
J. C. M. . . . Democritus reviviscens, etc. Pp. 457.
London, Ex officina Rogeri Danielis
1658
12°
Wing M255A; BL 1473. aa. 24.

[Cooper does not cite place and date of his edition.]

MAGNENUS, Joannes Chrysostomus (6)
J. C. M. . . . De Manna . . . Edition Ultima . . . repurgata.
Hagae-Comitis
1658
12°
BL 449. a. 18.

[Cooper does not cite place and date of his edition.]

MARBECKE, John (1)
The lives of holy Saints, Prophets, etc.
London, by R. N. for John Williams
1681
8° [4°, Wing]
Wing M565F; BL G. 19859.

MAXWELL, William (4)
Medicina Magnetica: or, the rare and wonderful Art of Curing by Sympathy; laid open by Aphorismes; proved in conclusions; and digested

into an easy method [translated from the Latin of William Maxwell by Christopher Irvine].
Edinburgh, C. Higgins [London, BL]
1656
8°
Wing I1053; BL E. 1578. (1.)

(4)
Dr. Meverell's Answers to the Lord Bacon's Questions, concerning the Compounding, Incorporating, Separation, Variation, and Restitution of Metals and Minerals in the Lord Bacon's Physiologial Remains [in *Baconiana. Or Certain Genuine Remains of Sr. Francis Bacon*].
London, by J. D. for Richard Chiswell
1679
8°
Wing B269; BL 535. b. 1.

OWEN, John (2)
Truth and Honesty in Plain English, or a Survey of the Libells and Pamphlets since the Dissolution of the Parliament.
London
1679
4°
Wing T3149

[Later edition of *Truth and Innocence vindicated* [by J. Owen] *in a survey of a Discourse* [by S. Parker] *concerning Ecclesiastical Polity*, etc. London, 1669, 8°; Wing O817; BL 4103. b. 54.]

P., H. (1,5)
Five Treatises of the Philosopher's Stone. Two of Alphonso, King of Portugall, as it was written with his own hand, . . . translated out of the Portuguez into English. One of J. Sawter . . . Another by F. Raudorff . . . Also a treatise of the names of the philosophers's stone, by W. Gratacolle. Translated into English. To which is added the Smaragdine Table. By the paines and care of H. P.
London, by Thomas Harper, to be sold by John Collins
1652 [1651 BL]
4°
Wing A2900; BL E. 654. (5.)

PARACELSUS (1,5,6)
Paracelsus his Archidoxes, disclosing the genuine way of making Quin-

tessences, Arcanums, Magisteries, Elixirs, &c.; Together with his books of Renovation & Restauration. Of the Tincture of the Philosophers. Of the Manual of the Philosophical Medicinal Stone. Of the Virtues of the Members. Of the Three Principles. And finally his seven books, of the Degrees and Composition of Receipts, and Natural Things. Faithfully and plainly Englished, and published by J. H. Oxon. 2 pt.
1660: London, for W. S. to be sold by T. Brewster, 24° Wing B3538
1661: London, for W. S. sold by Samuel Thomson, 8° Wing B3539; BL E. 2268.
1663: London, for Lodowick Lloyd, 8° Wing B3539A; BL 1489. k. 22.

PARACELSUS (5,6)
Paracelsus his Aurora, & Treasure of the Philosophers. As also the Water-stone of the Wise Men [by Ambrosius Siebmacher]; *describing the matter of, and manner how to attain the universal tincture. Faithfully Englished. And published by J. H. Oxon.* Pp. 229.
London, for Giles Calvert
1659
12°
Wing B3540; BL 8907. a. 22.

PARACELSUS (4)
Paracelsus his Mystical Philosophy written to the Athenians (see PINNELL, Henry, *Philosophy reformed and improved*)

PERSON, David (4)
Varieties, or, a Surveigh of rare and excellent matters . . . Digested into five Bookes, etc.
London, R. Badger for T. Alchorn
1636 [1635, STC, BL]
4°
STC 19781; BL 537. b. 2.

PETTUS, Sir John (4)
Fleta Minor. The laws of art and nature in knowing, judging, assaying . . . metals. In two parts. The first contains assays by L. Ercker in v. books . . . Translated into English. The second contains essays on metallick words alphabetically composed as a dictionary. By Sir J. Pettus. Illustrated with 44 sculptures.
London, by Tho. Dawks
1683
fol.
Wing P1906; BL L. 35/69.

PHILALETHA, Eirenaeus
Arcanum, or Secret of the Immortall Liquor Alkahest (see COOPER, William, *Collectanea Chymica*)

[PHILALETHA, Eirenaeus Philoponos] STARKEY, George (5,6)
The Marrow of Alchemy, being an experimental treatise [in verse], *discovering the . . . Mystery of the Philosophers Elixer, in Two Parts. By Eirenaeus Philoponos Philalethes.*
London
1654: [part one], by A. M. for Edw. Brewster, 8° Wing S5278; BL C. 117. a. 16.
1655: [part two], by R. I. for Edw. Brewster, 8° Wing S5279; BL C. 117. a. 16.

[PHILALETHA, Eyraeneus] STARKEY, George (1,5,6)
Secrets Reveal'd; or an open entrance to the Shut Palace of the King; containing, the greatest treasure in Chymistry never yet so plainly discovered. Composed by a most famous Englishman, styling himself Anonymus, or Eyraeneus Philaletha Cosmopolita . . . Published . . . by W. C., Esq., a true lover of Art and Nature [*with the Epistle Dedicatory of John Langius, Publisher of this author in Latin* (and the "Preface of John Langius to the Reader")]. Pp. 120.
London, W. Godbid for William Cooper
1669
8°
Wing S5288; BL 8905. aa. 19.

PHILALETHA, Philosophus (2,5,6)
Introitus Apertus ad occlusum Regis Palatium; auctore anonymo Philaletha Philosopho. In gratiam Artis Chymicae Filiorum nunc primum publicatus, curante J. Langio. Pp. 79.
Amsterdam, Jo. Janssonium
1667
8°
BL 1034. g. 16.

PHILALETHES, AEyrenaeus (1,4,5,6)
Opus tripartitum de Philosophorum Arcanis. Videlicet: 1, Enarratio methodica trium Gebri medicinarum; 2, Experimenta de praeparatione Mercurii Sophici; 3, Vade mecum philosophicum, sive breve manuductorium ad Campum Sophiae. Autore anonymo, sub nomine AEyrenaei Philalethes, natu Angli, habitatione Cosmopolitae.
London, for Wm. Cooper

1678
8°
Wing S5281; BL 1033. g. 18.

PHILALETHES, Eirenaeus (4,5,6)
Ripley Reviv'd: or, an Exposition upon Sir George Ripley's Hermetico-Poetical Works. Containing the plainest and most excellent Discoveries of the most hidden Secrets of the Ancient Philosophers, that were ever yet Published. Written by Eirenaeus Philalethes an Englishman, stiling himself Citizen of the World. [Edited by William Cooper.]
London, by T. Ratcliff & Nat. Thompson for Wm. Cooper
1678
8°
Wing S5286; BL 1034. f. 14.

["At the end Cooper has added in Latin: Porta Prima. De Calcinatione Philosophica; a chapter belonging to Philaletha's *Fons Chymicae Philosophiae*, which has been omitted by Burrius," Ferguson, II, 193.]

Contents:
1. *An exposition upon Ripley's Epistle to King Edward IV.*
2. *An exposition upon Ripley's Preface.*
3. *An exposition upon Ripley's First six Gates of the Compound of Alchymie.*
4. *Experiments for the preparation of Sophick Mercury.*
5. *A Breviary of Alchemy.*
6. *An exposition upon Ripley's vision.*

PHILALETHES, Eirenaeus (1)
Secret of the Immortal Liquor Alkahest (see COOPER, William, *Collectanea Chymica*)

PHILALETHES, Eirenaeus (2,5,6)
Three Tracts of the Great Medicine of Philosophers for humane and metalline bodies. I. Intituled, Ars metallorum metamorphoseos. II. Brevis manuductio ad rubinum coelestem. III. Fons chymicae philosophiae. All written in Latine by Eirenaeus Philalethes, Cosmopolita. Translated into English . . . by a Lover of Art, etc. [An English translation of the Latin *Tres Tractatus de Metallorum Transmutatione*, ed. by Martin Burrius, Amsterdam, 1668, 8°, J. Janssonium, BL 1033. g. 21.]
London, printed and sold by T. Sowle
1694
8°
Wing S5290; BL 1033. g. 19.

(4)
A Philosophical Riddle in Verse
(see HOUPREGHT, *Aurifontina Chymica*)

PINNELL, Henry (4)
Philosophy reformed and improved in four profound tractates: the 1, discovering the great and deep mysteries of nature, by O. Crollius; the other 3, discovering the wonderfull mysteries of the creation, by Paracelsus: being his philosophy to the Athenians. Both made English (and edited) *by H. Pinnell, etc.*
London, by M. S. for Lodowick Lloyd
1657
8°
Wing C7023; BL E. 1589. (1.)

PLATT, Hugh (3)
Sundrie new and artificiall Remedies against Famine. Written . . . uppon thoccasion [sic] *of the present dearth in the year 1569.*
London, P. Short
1569 [1596, BL and STC]
8° [4°, BL and STC]
STC 19996; BL C. 31. c. 10.

(4,5,6)
The Principles of the Chymists of London stated, with the Reasons of their dissent from the Colledge of Physicians. 2 pt. Pp. 128.
London, for the Authors
1676
8°
Wing P3497; BL 778. c. 38.

(4)
The Privy Seal of Secrets, discovering the first matter of the Philosophers
(see HOUPREGHT, *Aurifontina Chymica*)

RAY, John (4)
A Collection of English Words not generally used, with their significations and original, with catalogues of English birds and fishes; and an account of the preparing and refining such metals as are gotten in England. Pp. 178.
London, H. Bruges, for T. Burrell
1674

8°
Wing R388; BL 435. a. 26.

RIPLEY, George (3)
Bosome-Book of his Philosophical Accurations in the making of the Philosophers Mercury and Elixirs.
(see COOPER, William, *Collectanea Chemica*)

RIPLEY, George (4)
Treatise of Mercury (see HOUPREGHT, *Aurifontina Chymica*)

RODOCANACES, Constantinos (4)
A Discourse in the praise of antimonie and the vertues thereof.
London
1664
4°
Wing R1335; BL 7461. aaa. 63.

ST. DUNSTAN (5)
Of the Philosophers Stone (see COELSON, Lancelot, *Philosophia Maturata*, in *Catalogue of Chymicall Books*)

SALMON, William (5)
Synopsis medicinae or compendium of Astrological, Galenical and Chymical physick and philosophy, deducted from the Principles of Hermes and Hippocrates. In three books, etc.
London, by W. Godbid, for Richard Jones
1671
8°
Wing S454; BL 774. f. 2.

SCOT, Patrick (1)
The Tillage of Light, or, a true discoverie of the philosophicall Elixir, commonly called the Philosopher's Stone.
London, for W. Lee
1623°
8°
STC 21862; BL 1033. d. 9.

SEBER, Wolfgang (6)
Index Vocabulorum in Homeri non tantum Iliade atque Odyssea, sed caeteris etiam quotquot extant poematis, cum rerum . . . insigniorum anno-

tatione, catalogo item Graecolatino vocabulorum quibus apud Homerum, et ex eo in indice adjecta sunt epitheta. Studio W. Seberi editus. Pp. 557.
Heidelburg, in Bibliopolio Commeliniano
1604
4°
BL 652. c. 8.

[Cooper does not cite place and date of his edition.]

(4)
Secrets Disclos'd of the Philosophers Stone (see HOUPREGHT, *Aurifontina Chymica*)

SELDEN, John (6)
Ioannis Seldeni. De Synedriis & Praefecturis.
Londini, typis Jacobi Flesher, prostant apud Cornelium Bee
1650, 1655
4°
Wing S2425 or S2426

[Cooper does not cite place and date of his edition.]

SENDIVOGIUS, Michael (1)
A New Light of Alchymie: taken out of the fountaine of nature, and manuall experience. To which is added a treatise of sulphur . . . by M. Sandivogius [sic]. *Also nine books Of the Nature of Things, written by Paracelsus . . . Also a Chymicall dictionary explaining hard places and words in the writings of Paracelsus, and other obscure authors . . . Translated out of the Latin . . . by J. F., M. D.*
London, for Thomas Williams
1674
8° [12° , Wing]
Wing S2507; BL 1484. ee. 8.

SEILERUS, Wenceslaus (4)
(see BECHER, J. J., *Magnalia Naturae*)

SERGEANT, John (6)
A Vindication of the Doctrine contained in Pope Benedict XII, his bull, and in the General Councill of Florence, under Eugenius IIII concerning the state of departed souls. In answer to a certain letter, Printed and published against it, by an unknown author under this title A let-

ter in answer to the late dispensers of Pope Benedict XII his Bull; &c., wherein the progress of Master Whites lately minted Purgatory is laid open . . . By S. W. a Roman Catholick. Pp. 202.
Paris
1659
8°[12°, Wing]
Wing S2599; Cambridge

[Cooper does not cite place and date of his edition.]

SHIRLEY, Thomas (5)
A Philosophical Essay: declaring the probable causes, whence stones are produced in the greater world, with a search into the Origin of all bodies.
London, for William Cademan
1672
8°
Wing S3523; BL 954. a. 12.

SIMPSON, William (1)
A Discourse of the Sulphur-Bath at Knarsbrough in York-shire.
London, for William Cooper
1675
8°
Wing S3830; BL 1171. e. 12.

SIMPSON, William (4)
Philosophical Dialogues concerning the Principles of Natural Bodies, wherein the principles of the old and new philosophy are stated, and the new demonstrated, more agreeable to reason, etc.
London, by T. Hodgkin for Dorman Newman
1677
12°
Wing S3835; BL 536. a. 22.

SIMPSON, William (1,5,6)
Zymologia Physica, or a brief Philosophical Discourse of Fermentation, from a new Hypothesis of Acidum and Sulphur, whereby the Phoenomena of all natural Hot-Baths, the Generation of Minerals, the Production of many Acidulae or Spaw-waters, the grand appearances of Heat, Fire and Light throughout the triplicity of Natures Dominions, &c; with a Discourse of the Sulphur-Bath at Knaresbrough in Yorkshire.
London, by T. R. & N. T. for W. Cooper
1675 [1670, BL]

8°
Wing S3840; BL 1170. d. 39.

Smithaei Disticha Regum Angliae, in uno folio. **(6)**
[poss. William Smith's *Carmen triumphale, or Englands triumph for her restored libertie. With White-Halls speech to her royal master, Charles the second King, etc.*]
London
1660
4°
Wing S4273

SPENSER, Edmund (5,6)
Spencer's [sic] *View of Ireland, fol.* [Probably the edition included in Sir James Ware, *Two Histories of Ireland.*]
Dublin, Society of Stationers
1633
fol.
STC 25067; BL 601. l. 1.

STARKEY, George (3)
Admirable Efficacy . . . of the true Oyl of Sulphur-Vive (see COOPER, William, *Collectanea Chemica*)

STARKEY, George (1,5)
Liquor Alchahest, or a Discourse of that Immortal Dissolvent of Paracelsus and Helmont, it being one of those two wonders of Art and Nature, which radically dissolves all Animals, Vegetals, and Minerals into their principles, without being in the least altered, either in weight or activity after a thousand dissolutions, &c. Pp. 55.
London, by T. R. & N. T., for W. Cademan
1675
8°
Wing S5277; BL 717. a. 50.

STARKEY, George [PHILALETHES, Eirenaeus Philoponos] (2)
Marrow of Alchemy, being an experimental treatise (in verse), discovering the . . . Mystery of the Philosophers Elixer. 2 pt.
London
1654: by A. M. for Edw. Brewster, 8° Wing S5278; BL C. 117. a. 16.
1655: by R. I. for Edw. Brewster, 8° Wing S5279; BL C. 117. a. 16.

STARKEY, George (5)
Pyrotechny asserted and illustrated, to be the surest and safest means for arts triumph over nature's infirmities, being a full and free discovery of the medicinal mysteries studiously concealed by all artists, and only discoverable by fire: With an Appendix concerning the nature, preparation, and virtue of the great Arcana.
London, by R. Daniel, for Samuel Thomson
1658
8°
Wing S5284; BL 1035. a. 13.

STRADA, Famianus (6)
Strada's Musical Duel . . . much enlarg'd, in English, by the addition of several traverses between the Harper and Nightingale; together with a more particular account of the contest.
London, by J. W. for William Gilbert
1671
4°
Wing S5780; BL 11405. bbb. 37. (5.)

(4)
A strange Letter concerning an Adept (see HOUPREGHT, *Aurifontina Chymica*)

(4)
SYNESIUS the Greek Abbot, Of the Philosophers Stone (see VALENTINE, *Triumphant Chariot of Antimony*, 1678 edition; in *Catalogue of Chymicall Books*)

THOMSON, George (5,6)
Epilogismi Chymici, observationes nec non remedia Hermetica longa in Arte Hiatrica exercitatione constabilita, Item Essentiae nostrae stomachicae vires insignes Medicae explicantur, ejusque materia, Modus ac Methodus praeparationis ad Galeno-Chymicorum Elenchum . . . describuntur.
London, for Nathaniel Crouch
1673
8°
Wing T1022; BL 1143. a. 29.

THOMSON, George (1,5,6)
μεθοδος ιατροχυμικη; or, the direct method of curing chymically. Wher-

ein is conteined the original matter, and principal agent of all natural bodies, also the efficient and material cause of Diseases in general; . . . with the Art of Midwifery chymically asserted . . . A Description of the Sanative Virtues of our Stomach Essence, etc.
London, for B. Billingsley, & S. Crouch
1675
8°
Wing T1029; BL 1163. b. 54.

(4)
The Treasure of Treasures (see HOUPREGHT, *Aurifontina Chymica*)

(4)
A Treatise of the Blessed Manna of the Philosophers
(see HOUPREGHT, *Aurifontina Chymica*)

(4)
TREVISAN, Bernhard, Earl of (see BERNHARD)

(4)
Tumba Semiramidis (see COOPER, William, *Collectanea Chymica*)

USHER, James (6)
Gotteschalci, et Praedestinationae controversiae ab eo motae, historia: una cum duplice ejusdem confessione, nunc primum in lucem edita, etc.
Dublin, Societatis Bibliopolarum
1631
STC 24550; BL 4255. bb. 74.

USHER, James (6)
Veterum epistolarum Hibernicarum sylloge; quae partim ab Hibernis, partim ad Hibernos, partim de Hibernis vel rebus Hibernicis sunt conscriptae. J. Usserius . . . collegit at recensuit.
Dublin, Societatis Bibliopolarum
1632
4°
STC 24557; BL 601. d. 3. (1.)

VAUGHAN, Thomas [PHILALETHES, Eugenius] (1,5)
Aula Lucis, or, the House of Light; A Discourse written in the year 1651. By S. N. A modern speculator.
London, for William Leake

1652
8°
Wing V144; BL 1033. d. 14.

WARE, Sir James (6)
De Scriptoribus Hiberniae libri duo. Prior continet Scriptores in Hibernia natos. Posterior, Scriptores alios qui in Hibernia munera aliqua obierunt.
Dublin, Societatis Bibliopolarum
1639
4°
STC 25066; BL 601. d. 3. (2.)

WEBSTER, John (2)
Metallographia, or an history of metals, with Signs of Ores, and minerals, before and after diging, the causes and manner, of their Generations, their kinds, sorts, and differences, with a description of new metals, and Semi metals, and other things pertaining to mineral knowledge, also of Vegitability, of mystical Chymistry, of the Philosophers Gold and Mercury, of the Liquor Alkahest, Aurum Potabile, and such like.
London, A. C. for Walter Kettilby
1671
4°
Wing W1231; BL 990. e. 1.

WILLIS, Timothy (5,6)
The Search of Causes. Containing a Theophysicall investigation of the possibilitie of transmutatorie Alchemie.
London, J. Leggatt
1616
8°
STC 25754; BL 1036. a. 13. (2.)

Appendix II
William Cooper and the Works of Eirenaeus Philalethes

Cooper's interest in current works on alchemy and chemical medicine is nowhere more evident than in his promotion and publication of treatises that passed, albeit with considerable confusion and uncertainty, under the pseudonymous authorship of Eirenaeus Philalethes. The problem of establishing the true identity of this adept—compounded by a proliferation of similar pseudonyms in the seventeenth century, e.g. Eugenius Philalethes (Thomas Vaughan), Alazonomastix Philalethes (Henry More), Eirenaeus Philoponos Philalethes (probably George Starkey)—has resisted definitive solution despite numerous attempts by modern scholars and bibliographers. Most frequently, Eirenaeus Philalethes has been closely associated with the Bermuda-born, Anglo-American chemist and physician George Starkey (or Stirk), who graduated from Harvard in 1646, emigrated to England around 1650, became well connected in London chemical and medical circles, and quite possibly died in the Great Plague of 1665. The exact nature of the connection between Starkey and Philalethes has been interpreted variously, however. A. E. Waite, for example, speaks of Starkey as one "who, owing his initiation to [Philalethes], may be considered his philosophical son, but not his *alter ego*."[1] John Ferguson notes only the obvious fact that Philalethes' writings have often been attributed to Starkey.[2] R. S. Wilkinson, whose investigation of the problems relating to the identities of and relationship between Philalethes and Starkey is most recent and illuminating, reaches two very different conclusions. He argued in 1964 that the true author of the "Philalethean" tracts that Starkey took to England and circulated there was John Winthrop, Jr. (1606–1676) of Massachusetts and Connecticut, the owner of a very extensive collection of alchemical books.[3] In 1972 and again in 1973, however, Wilkinson reversed his earlier position, arguing, on the basis of circumstantial evidence, for Starkey's authorship of the Philalethean materials.[4]

While questions of the merit of these theories of authorship are outside the scope of this discussion, it is important to note that Cooper, who published works by both Starkey (e.g. the 1684 edition of the *Liquor Alchahest*) and Philalethes, consistently sees the two as having separate identities. Both Wilkinson and Macphail present evidence to suggest the opposite, citing Cooper's entry for *The Marrow of Alchemy* in the 1673 version of the *Catalogue of Chymical Books*:

> Eir. Phil. Philalethes, *alias* Geo. Starkies *Marrow of Alchimy, being an experimental Treatise of the secret and most hidden mystery of the Philosophers Elixir,* Lond. 1654. 8.[5]

However, both Wilkinson and Macphail overlook the fact that in the 1675 edition of the *Catalogue . . . In Three Parts*, Cooper issues a correction of this entry in the form of "An Advertisement," which immediately precedes the beginning of the booklist:

> *In the First Part of the* Catalogue [i.e., 1673 and the First Part of 1675] *these three mistakes were committed, viz.* Dr. Croon's *Letter instead of* P. M's; Geo. Starkie *was put down for the Author of the* Marrow of Alchymie, whereas *he was but the Publisher: his Name was likewise put to the* Marrow of Chymical Physick, *instead of* Will. Thrasher, *who is the right Author thereof.*[6]

Since the First Part of the 1675 *Catalogue* is identical to the 1673 edition, this correction does not appear in the entry included in this part; however, in the Second Part of 1675, Cooper has re-set the entry and deleted "alias Geo. Starkies."

All other entries for works by Starkey and Philalethes in the 1675 *Catalogue* and the 1688 *Continuation or Appendix* have been segregated scrupulously. Furthermore, there is no evidence in any of the editions of the *Catalogue* that he confused the identity of Eirenaeus Philalethes with Eugenius or any other "Philalethes" whose works were then circulating. Thus, having corrected the "alias Geo. Starkies" mistake at the time the Second Part of the 1675 *Catalogue* was being set for printing Cooper appears to have removed from his own mind any possible doubts and confusions about authorship. For him, George Starkey was not the author of the *Marrow of Alchemy*, and he and Eirenaeus Philalethes were not the same persons.

As revealed in the entries of the *Catalogue of Chymicall Books* and in those supplementary titles from booklists included in Appendix I, throughout his publishing career, Cooper persisted in keeping his readers apprised of available works by Philalethes. In the period 1677–78, however, this determination became even stronger. We see this, for example, in the one-page list of "*Books . . . Printed lately for* William Cooper *at the* Pellican *in* Little-Britain, *London*," included in *A Breviary of Alchemy* (Wing S5271), which later appeared in the collection, *Ripley Reviv'd.* Of the twelve titles included in this short list, five are those of the individual sections of *The Philosophical Epitaph*, including the *Catalogue*; the sixth is William Simpson's *Zymologia Physica, or a brief Philosophical Discourse of Fermentation* (London, 1675); and the remaining six are Philalethean works either recently published (*Secrets Reveal'd*, the *Exposition upon Sir G. Ripley's Epistle to King Edward IV*, and *A Commentary upon Sir G. Ripley's Recapitulation*, i.e., *A Breviary of Alchemy*) or currently being printed (the *Exposition upon Sir G. Ripley's Vision*, his *Preface*, and upon the *First six Gates of the Compound of Alchymie*).[7]

A far more comprehensive and valuable list, transcribed below, is the one that Cooper included with his edition of *Ripley Reviv'd* of 1678 (Wing S5286), which is a gathering of the five aforementioned treatises on Ripley, in addition to the *Experiments for the Preparation of the Sophic Mercury.* Of special value for its information on the relationship between Starkey and Philalethes and for the insights it provides into Cooper's keen interest in publishing Philalethean material, is the *Advertisement* that immediately precedes the two-page listing of the adept's works. Cooper begins by stating that Philalethes' motive in writing his "many Excellent Pieces" was not personal fame and fortune, but the desire to contribute to mankind through helping "*all sincere searchers of this secret Art.*" The *Advertisement* is primarily interesting, however, for the light it throws on Starkey's role in the transmission of Philalethes' works and on Cooper's own determination to locate and eventually publish the missing Philalethean commentaries on the last half of Sir George Ripley's *Compound of Alchymy.* An accurate and trustworthy authority in matters of books and authors, Cooper again speaks with complete confidence as to the separateness of the identities of Starkey and Philalethes:

> Seeing therefore, that it was the Authors own desire to benefit the World by his Labours, and that he gave his consent to Mr. *Starkey* for Printing his Pieces, as appears in his *Preface to the Marrow of Alchimy*; I know no reason wherefore his Writings should lie conceal'd any longer: And great pity it was that Mr. *Starkey* did separate this Author's *Commentarie upon Sir* George Ripley's 12 *Gates*, which he did as I was informed by one unto whom he gave

the very Book from which he confessed he had cut the last Six Gates; the Person demanding the reason wherefore he cut them in sunder: he answered, that the World was unworthy of them; which nevertheless he promis'd to give that Person a Transcript of, but did not, which is the reason that they cannot yet be found; the loss of which is very much lamented. Wherefore if any Gentleman hath them by him, or any other piece of this Authors, It is humbly desired that they will send them to the *Pellican* in *Little Britain, London*, that they may be Printed with the first *Six Gates*, which are now in the Press: And that I may not be wanting to contribute what I can for the discovery of this Author's Works, I here make bold to present the Reader with a Catalogue of such Pieces as are noted to be writ by this Author under the disguised name of *Æyraeneus Philalethes*, part whereof are set down by Mr. *Starkey* in his Preface aforenamed, and part are mentioned by the Author himself, with several others, which he wrote (as he saith) for his own recreation, and afterwards burn'd; which Author is acknowledged by all hands to be an English-man, and an Adept & supposed to be yet living, and travelling, and about the age of 55 years, but his Name is not certainly known.[8]

These Books in this *Catalogue* were written
by *Eirenaeus Philalethes*,
whereof these 15. following are Printed[9]

1. *Introitus apertus ad occlusum Regis Palatium*, Amst. 1667. This is Re. printed in *Germany*, with the Collection of Books called *Musaeum Hermeticum* of the Edition 1677 . . . in 4°[10]

2. ________*Idem* in English, called *Secrets Reveal'd*, Printed at *London* 1669. in 8° being much more perfect than the Latine Editions.[11]

3. *The Marrow of Alchemy*, in two Poems or Parts, in English Verse, *Lond*. 1654. & 1655.[12]

(Amst. 1668. in 8° These are likewise in the
aforesaid Edition of the *Musaeum Hermeticum*.)[13]

4. *Ars Metallorum Metamorphoseos,*

5. *Brevis manuductio ad Rubinum Coelestem,*

6. *Fons Chymicae Philosophiae,*

(Lond. 1678, in 8°)[14]

7. *Methodica Enarratio trium Gebri Medicinarum,*

8. *Vade-Mecum Philosophicum, sive breve manuductorium ad campum Sophiae,*

9. *Experimenta de praeparatione Mercurii Sophici,*

(Lond. 1678, in 8°)[15]

10. A Commentary or Exposition upon Sir *G. Ripley*'s Epistle to *Edw.* IV. King of *England,*

11. ________*Idem* upon Sir *G. Ripley*'s Preface to his Compound of Alchymy,

12. ________*Idem* upon the first six Gates of his Compound of Alchymy,

13. ________*Idem* upon the Recapitulation of his Compound of Alchymy,

14. ________*Idem* upon his Vision,

15. Experiments for the Preparation of the *Sophick Mercury,*

(These 13. following he wrote, but we cannot as yet find where the Copies are.)

1. A Comment. or Exposition upon the last six Gates of Sir *G. Ripley*'s Compound of Alchymy.

2. ________*Idem* upon Sir *G. Ripley*'s Erroneous Experiments.

3. ________*Idem* upon Sir *G. Ripley*'s Wheel.

4. ________*Idem* upon *Arnold's Ultimum Testamentum.*

5. *Opus Elixeris Aurifici & Argentifici.*

6. *Brevis via ad vitam longum*, or Alchymy Triumphing.

7. *Cabala Sapientum*, or an Exposition upon the Hieroglyphicks of the Magi.

8. *Elenchus Errorum in Arte Chymica deviantium.*

9. *Elenchus Authorum potissimorum in Arte Chymica.*

10. An *Enchiridion* of Experiments, together with a Diurnal of Meditations, in which were many Philosophical Receipts, declaring the whole Secret; with an *Ænigma* at the end.

11. *Analysis Operis.*

12. A *Clavis* to his Works.

13. Comments or Expositions upon *Flammel, Artephius*, and *Sendivogius*. But these three are rather Quæried, then affirmed to be wrote by this Author.

Notes

[1] Arthur E. Waite, *Alchemists through the Ages* (1888; rpt. [of *Lives of Alchemystical Philosophers*] Blauvelt, N.Y.: Rudolf Steiner, 1970), p. 190.

[2] John Ferguson, *Bibliotheca Chemica* (1906; rpt. London: Derek Verschoyle, 1954), II, 194.

[3] See Ronald S. Wilkinson, "The Problem of the Identity of Eirenaeus Philalethes," *Ambix*, 12 (February, 1964), 24–43. For a thorough discussion of Starkey's career, see Wilkinson, "George Starkey, Physician and Alchemist," *Ambix*, 11 (October, 1963), 121–52; and on Winthrop and the contents of his library see Wilkinson, "The Alchemical Library of John Winthrop, Jr. (1606–1676) and his Descendants in Colonial America," *Ambix*, 11 (February, 1963), 33–51 and 13 (October, 1966), 139–86.

[4] "Further Thoughts on the Identity of 'Eirenaeus Philalethes,'" *Ambix*, 19 (November, 1972), 204–08 and "Some Bibliographical Puzzles Concerning George Starkey," *Ambix*, 20 (November, 1973), 235–44.

[5] William Cooper, *A Catalogue of Chymicall Books which have been written Originally, or Translated into English* in *The Philosophical Epitaph of W. C. Esquire* (London, 1673), sig. Q4^v. See Wilkinson, "The Problem of the Identity of Eirenaeus Philalethes," p. 26; and Ian Macphail, *Alchemy and the Occult* (New Haven: Yale University Library, 1968), II, 481.

[6] William Cooper, *A Catalogue of Chymicall Books. In Three parts* (London, 1675), printed immediately following the *Bibliopola Lectori*.

[7] Eirenaeus Philalethes, *A Breviary of Alchemy* (London, 1678), p. [29].

[8] Printed following *An Exposition Upon Sir George Ripley's Epistle to King Edward IV*, [pp. 45–46], in *Ripley Reviv'd* (London, 1678). An identical printing of the *Advertisement* and the following list of works by Philalethes appeared in the separate edition of *An Exposition* printed by Cooper in 1677.

[9] The following transcription, with the format slightly modified from the original, appears on pp. 47–[48] of *Ripley Reviv'd* (London, 1678), following the tract entitled *An Exposition Upon Sir George Ripley's Epistle to King Edward IV*. In the same year, Cooper printed a very similar list in Latin, the *Catalogus Librorum editorum Authore Æyrenaeo Philalethe Cosmopolita*, in the *Enerratio Methodica* (London, 1678), [pp. 223–24], (Wing S5273).

[10] For full citation, see Appendix I. The reprint appears in the *Musaeum Hermeticum reformatum et amplificatum, omnes Sopho-Spagyricae artis discipulos . . . erudiens, quo pacto . . . lapidis philosophici medicina . . . inveniri et haberi queat. Continens tractatus chimicos xxi. In gratiam filiorum doctrinae, quibus Germanicum idioma ignotum est, Latina lingua ornatum.* Francofurti, 1677.

[11] See *Catalogue*, entry no. 293.

[12] See *Catalogue*, entry no. 294.

[13] For full citation, see Appendix I: PHILALETHES, Eirenaeus, *Three Tracts of the Great Medicine*.

[14] For full citation, see Appendix I: PHILALETHES, Aeyrenaeus, *Opus Tripartitum*.

[15] See *Catalogue*, entry no. 295.

Subject Index

The following numbers refer to the entries in the *Catalogue of Chymicall Books.*